Comparative Pathobiology

Volume 7

PATHOGENS OF INVERTEBRATES

Application in Biological Control and Transmission Mechanisms

Comparative Pathobiology

Volume 1: Biology of the Microsporidia
Volume 2: Systematics of the Microsporidia
Volume 3: Invertebrate Immune Responses
Volume 4: Invertebrate Models for Biomedical Research
Volume 5: Structure of Membranes and Receptors
Volume 6: Invertebrate Blood: Cells and Serum Factors
Volume 7: Pathogens of Invertebrates: Application in Biological Control and Transmission Mechanisms

Comparative Pathobiology

Volume 7

PATHOGENS OF INVERTEBRATES
Application in Biological Control and Transmission Mechanisms

Edited by Thomas C. Cheng
Medical University of South Carolina
Charleston, South Carolina

Plenum Press • New York and London

Library of Congress Cataloging in Publication Data

Main entry under title:

Pathogens of invertebrates.

(Comparative pathobiology; v. 7)
"Proceedings of the thirteenth annual meeting of the Society for Invertebrate Pathology, held July 26–August 2, 1980, in Seattle, Washington"—T.p. verso.
Includes bibliographical references and index.
1. Insects—Diseases—Congresses. 2. Invertebrates—Diseases—Congresses. 3. Micro-organisms, Pathogenic—Congresses. 4. Insect control-Biological control—Congresses. I. Cheng, Thomas Clement. II. Society for Invertebrate Pathology. Meeting (13th: 1980: Seattle, Wash.) III. Series.
SB942.P38 1984 632'.62 84-6773

ISBN 978-1-4615-9390-4 ISBN 978-1-4615-9388-1 (eBook)
DOI 10.1007/978-1-4615-9388-1

Proceedings of the Thirteenth annual meeting of the Society for Invertebrate Pathology, held July 26–August 2, 1980, in Seattle, Washington

Softcover reprint of the hardcover 1st edition 1984

A Division of Plenum Publishing Corporation
233 Spring Street, New York, N.Y. 10013

PREFACE

Invertebrate pathology, like medical and veterinary pathology, for many years has been spearheaded by practical applications although in more recent times many investigators have elected to focus their attention on basic mechanisms and the elucidation of basic phenomena. Although Elie Metachnikoff and Louis Pasteur may be considered the forerunners of invertebrate pathology, in modern times the late Edward A. Steinhaus and the late Arthur M. Heimpel, among others, must be considered the principal disciples. Consequently, in recent years several symposia have been organized in honor of the memory of Steinhaus and Heimpel. When the proceedings of these occasions were examined and reviewed, it was decided that these could naturally be considered chapters of a single volume of *Comparative Pathobiology* under the subtitle selected.

We wish to note that the chapters devoted to various aspects of *Bacillus thuringiensis* were originally presented at the thirteenth annual meeting of the Society for Invertebrate Pathology held in Seattle, Washington, on July 26 - August 2, 1980, under the title of the "Edward A. Steinhaus Memorial Symposium". This includes the contribution by Dr. Robert M. Faust on the professional contributions of Dr. Arthur M. Heimpel. Art, as he was known to his friends, was a founding member of the Society and later served as its president. The circle of his professional associates was international. Consequently, it was not surprising that we received many requests from insect pathologists all over the world to publish Bob Faust's tribute so that a permanent record of Art Heimpel's career will be available.

There can be no doubt that the discovery of *Bacillus thuringiensis* and subsequent development of this bacterium (and the many varieties and serotypes) into the major biological control agent against certain noxious and destructive insects collectively stand as a major landmark in the practical application of knowledge gained from studying invertebrate pathology. Consequently, the chapters included herein on *B. thuringiensis* also represent permanent records of this major saga.

In the area of molluscan pathology, specifically the diseases of oysters, the haplosporidan protozoans represent a major group of pathogens. The literature pertaining to these pathogens is spread widely. Consequently, the contribution by Dr. Jay D. Andrews represents a valuable attempt at summerizing what is known.

Finally, the contribution by Dr. P. E. M. Fine on verticle transmission of pathogens of invertebrates, in our opinion, stands as an important critique that will undoubtedly be repeatedly cited by invertebrate pathologists and epizootiologists (= epidemiologists).

Other volumes of *Comparative Pathobiology* are in preparation or being planned. It has been most gratifying to know that this series is being widely read and appreciated.

Thomas C. Cheng
Charleston, South Carolina

CONTENTS

ARTHUR M. HEIMPEL: HIS SERVICE AND CONTRIBUTIONS TO THE PATHOBIOLOGY OF *BACILLUS THURINGIENSIS* AND OTHER ENTOMOPATHOGENS

Robert M. Faust

Insect Pathology Laboratory
Agriculture Research
Science and Education Administration
U.S. Department of Agriculture
Beltsville, Maryland

---

[1]Mention of a proprietary product or a company name in this paper does not constitute an endorsement by the U.S. Department of Agriculture.

I. INTRODUCTION

Arthur MacLeod Heimpel, an outstanding leader in insect pathology, died on 10 November, 1979, at his home in Beltsville, Maryland. He was born in Baie d'Urfe, Quebec, Canada, on 27 June, 1923, the son of Dr. Louis G. Heimpel, a Professor at MacDonald College, and his wife, nee Blanche M. MacLeod. He graduated from MacDonald College, where he received his secondary school education, and completed military duty overseas with the Canadian Army (1943-1946). He did his undergraduate work at Queen's University, Kingston, Ontario (B.A., 1948). His graduate years were spent at Queen's University (M.A., 1949; Ph.D., 1954) and at the University of California, Berkeley. He was awarded the first Ph.D. in Biological Sciences to be granted by Queen's University. While at Queen's University, he married Elma Rae Lyle on 26 July, 1947. He is survived by his wife and 5 sons, Gordon, Bruce, Michael, Ernest, and David.

During his studies at Queen's University, he worked under G. B. Reed, well known for his work on bacterial toxins. At Berkeley he studied under Edward A. Steinhaus, who became one of his most influencial mentors. In his Presidential Address delivered at the 7th annual meeting of the Society for Invertebrate Pathology in 1974, he stated in retrospect, "I became irrevocably committed to the field of insect pathology in 1947 when Dr. A. S. West of Queen's University, Kingston, Ontario, Canada, handed me a book entitled *Insect Microbiology* by E. A. Steinhaus, and suggested that I read it. In 1949, I was in Berkeley, California, buying one of the first copies of Steinhaus' book entitled *Insect Pathology*..." Presumably, this reading and the influence of Professor Steinhaus gave the final impetus to his principal interest and direction in general insect pathology and microbial control.

Dr. Heimpel began to apply his interest during a 13-year period (1948-1961) spent first as a student and then as a research officer at the Insect Pathology Institute, Canadian Department of Agriculture, Ontario, Canada. The remainder of his career centered at the Insect Pathology Research Laboratory, U.S. Department of Agriculture, Agricultural Research Center, Beltsville, Maryland, first as Principal Insect Pathologist (1961-1970) and then as Laboratory Chief (1970-1979). There he was fascinated with the problems of mode of action, classification, commercial production, and safety of entomopathogenic bacteria and viruses.

On 19 August, 1968, he became a naturalized citizen of the United States, although he remained proud of his heritage long after his departure from Canada. We who were fortunate to know

Art, as he was known to his many friends, intimately remember a warm-hearted individual who always had the time to help colleagues and friends with their problems out his great store of knowledge. At the same time he demanded from his laboratory colleagues thoroughness, accuracy, and efficiency. He applied authority with grace and honest humility, nurturing an unpretentious, cheerful availability and generous gift of time and talent; when necessary, he motivated his people and events to their proper destiny.

II. SERVICES TO THE SCIENTIFIC COMMUNITY

The distinguished career of Arthur M. Heimpel may be divided into several major fields of endeavor that readily intermesh and cannot be separated chronologically. Whatever his involvement, however, he always continued his efforts to improve the science and the profession of insect pathology. Some degree of chronological description is important however, because it illustrates the concept that, in scientific work, one endeavor leads to another. Nonetheless, it is useful to remember that during Heimpel's 19 years at the Insect Pathology Laboratory in Beltsville, he was also a part-time graduate faculty member (1966-1979) and lecturer in the College of Agriculture, University of Maryland, where for several years he taught the first course in insect pathology. He also advised and served on the thesis committees of a number of graduate students in microbiology and entomology during this period. Thus, teaching, whether formal or informal, was a significant, continuing activity.

Between 1948 and 1961, Dr. Heimpel actively collaborated in laboratory and field studies with students and colleagues at the Forest Insect Laboratory in Sault Ste. Marie, Ontario, Canada. In 1960 he was involved with the first aerial spraying and testing of a preparation of *Bacillus thuringiensis* against the spruce budworm and black-headed budworm. These investigations were published in 26 articles concerned mainly with the chemistry, physiology, and pathology of entomopathogenic bacteria, especially *Bacillus cereus, Serratia marcescens,* and the crystalliferous bacteria. An off-shoot of some of these research studies was his text on insect diseases caused by spore-forming bacteria, published as a chapter in E. A. Steinhaus' book *Insect Pathology: An Advanced Treatise* in 1963. During this period of intensive research, study, and publishing, Dr. Heimpel became an expert bibliophile and seeker of references to published works. Perhaps because of this interest he became a member of the Editorial Board of the *Journal of Invertebrate Pathology* early in the 1960s. He also wrote more than 50 other articles (see Appendix) on subjects that concerned the history and status of insect pathology, actions of insect pathogens in their hosts, safety and commercialization of microbial agents, and the place and importance of insect pathology at the national and international levels.

His appointment as Chief of the Insect Pathology Laboratory in Beltsville (1961-1979) was noteworthy because it encompassed a broad challenge and allowed him the much desired opportunity to influence and participate in the development of research on insect pathology in the United States at the national level. The laboratory was first chartered as the Insect Pathology Pioneering Research Laboratory (1961-1970); its name was later changed to the Insect Pathology Laboratory (1970-present). The laboratory was charged with the responsibility of a long range, basic research program on the pathology of insects, with particular reference to microbiological organisms that cause diseases. The charter authorized studies involving investigations of the physiology, nutrition, genetic variability, and mode of action of bacteria, protozoa, nematodes, viruses, and fungi that were pathogenic to insects, which Dr. Heimpel often listed as his major interests. These studies included methods for propagation of pathogens and changes in the normal anatomy, physiology, and metabolism of infected hosts. Contemporary findings at that time suggested that the viruses and spore-forming bacteria were the most likely of the microorganisms to be practical in the control of pest insects; accordingly, Dr. Heimpel particularly emphasized fundamental studies of these organisms in the research program of the laboratory.

Under his leadership, the laboratory accelerated its involvement in the many facets of insect pathology, which included: searches for pathogens occurring naturally in insect populations and the development of methods for the identification of microorganisms encountered; methods for isolating pathogens from nonpathogenic microorganisms associated with them and for propagating them as a means of further study under controlled conditions; detailed investigations of the precise nature and manner in which pathogens invade and kill the insect, whether through invasive mechanisms or toxins; investigations of the comparative pathology and host range of various pathogenic microorganisms, the factors responsible for relative or absolute host specificity, and the factors limiting their rate of development or effectiveness under various conditions; studies on methods of disseminating insect pathogens so as to induce epizootics in natural insect populations; and studies on the taxonomy and classification of microorganisms, to properly describe them according to the international rules of nomenclature for microorganisms and to contribute to an understanding of the correlations between the phylogeny of organisms and the hosts with which they are associated.

During the last several years with the Insect Pathology Laboratory, he also was actively involved in the development of protocols for the various tests that were prerequisites to registration of microbial agents. He undertook the compilation of two voluminous petitions for submission to the U.S. Environmental Protection Agency. These documents were (1) Petition for Exemption from the Requirement of a Tolerance for Usage of the Nuclear Polyhedrosis Virus from the Alfalfa Looper, *Autographa californica*, on Lettuce and Cabbage (2 volumes); and (2) Petition for Exemption from the Requirement of a Tolerance for Usage of the Milky Spore Disease Bacterium, *Bacillus popilliae*, on pastures.

In addition to his substantial research and administrative burdens, he worked actively in the programs of the Society for Invertebrate Pathology from its inception in 1967 through 1979. He was a member of the Society's Founding Committee and a charter member. He served as its Founding Secretary-Treasurer (1967-1968), as a Trustee (1969-1972), as its Vice-President (1972-1974), then as its President (1974-1976).

It is difficult to list all of Arthur Heimpel's activities and the positions he held in scientific societies, organizations, and committees; there are so many of them that only some of the more important can be cited. Dr. Heimpel is cited in *Who's Who Biographies of Living Notables*. The following is an abbreviated selection made from a list of Dr. Heimpel's numerous activities and accomplishments.

Professional Affiliations and Committees

American Association for the Advancement of Science

Participant, Symposium on Alternative Methods of Insect Control, Montreal, 1964

American Chemical Society

Participant, Symposium on Pesticide Control and Eradication, Atlantic City, 1962

American Institute of Biological Sciences

Elected to Membership, 1968

Governing Board Representative of the Society for Invertebrate Pathology, 1974-1976

American Society for Microbiology

Elected to Membership, 1961

Chairman, Session on Spore Formers, Atlantic City, 1965

Participant, First International Congress of Comparative Virology, Quebec, 1969

Participant, Symposium on Microbial Insecticides, Washington, D.C., 1964

Participant, Symposium on Microbial Control, Montreal, 1964

Canadian Society for Microbiology

Elected to Membership, 1958

Commission Internationale de Lutte Biologique

Member, International Committee of the Insect Pathology Colloquia for Insect Control, 1962-1979

Entomological Society of America (ESA)

Elected to Membership, 1964

Chairman, Insect Pathology Symposium on Specificity of Insect Viruses, New York, 1967

Participant, XIIth International Congress of Entomology, London, 1964

Participant, XIVth International Congress of Entomology, Symposium on Microbial Control, Canberra, 1972

Entomological Society of Canada

Elected to Membership, 1948

Participant, Joint Meeting of the Entomological Societies of America, Canada and Ontario, Symposium on Biological Alternatives to Chemical Control, Detroit, 1969

Entomological Society of Ontario

Director, 1958-1959

Environmental Protection Agency

Participants, Symposium on Impact of Use of Microbial Organisms on Aquatic Environments, Pensacola, 1974

Technical Advisor, Safety of Microbial Control Agents, 1973-1979

Food and Drug Administration

Technical Advisor, Virus Safety Protocol Committee, 1969

International Scientific Committee for Colloquia on Insect Pathology

Member, 1962-1973

International Union of Biological Sciences

Chairman, Commission on Invertebrate Pathology, 1974-1976

International Virus Standardization Committee

Chairman, 1966-1967

Joint U.S.-Japan Seminar on Microbial Control of Insect Pests

Participant, Proceedings on Progress in Developing Insect Viruses as Microbial Control Agents, Fukuoka, 1967

Journal of Insect Pathology

Editorial Board, 1961-1964

National Aeronautics and Space Administration (NASA)

Consultant, NASA Johnson Space Center, Houston, Apollo 11 and 16, 1969-1974

National Academy of Sciences, National Research Council

Member, Subcommittee on Insect Pests

National Science Foundation

Commission on Exchange of Science and Technology with the USSR, Member, Subcommittee on Production of Materials by Microbial Means, 1974-1975

New York Academy of Sciences

Participant, Symposium on Regulation of Insect Populations by Microorganisms, New York, 1972

Society for Industrial Microbiology

Chairman, Symposium on Microbial Insecticides, Corvallis, Oregon, 1962

Society for Invertebrate Pathology

Founding Committee, Secretary-Treasurer, 1967-1968

Founding Member, 1967

Trustee, 1969-1972

Vice-President, 1974-1976

IVth International Colloquium on Insect Pathology, College Park, Member Scientific Committee and Organizing Committee, Secretary-Treasurer, 1970

Participant, IIIrd International Colloquium on Insect Pathology, Wageninger, The Netherlands, 1966

U.S. Department of Agriculture (USDA)

USDA North Central Region 72 Committee on Funds for Land-Grant College Research, Advisor, 1963-1979

National Technical Advisor, Insect Pathology and Microbial Control, 1972-1979

Member, Planning Group for Dedication of the Beltsville Agricultural Research Center Bioscience Building, 1972-1973

Participant, Ist Beltsville Symposium in Agricultural Research, Virology in Agriculture, Beltsville, 1976

Member, Program Committee, Vth Beltsville Symposium in Agricultural Research, Biological Control in Crop Production, Beltsville, 1979

Joint Project of the European Parasite Laboratory and Plant Protection Institute, Leader, Survey for Noctuid Insect Pathogens, Austria, 1978

Public Law (PL) 480 Projects (Cooperating Scientist)
Studies on Microbiology and Pathology of Insect Pests of Crop Plants. University of Agricultural Sciences, Bangalore, Mysore, India, 1965-1968

Studies on Interactions of Various Pathogens in One Insect Host (Cutworms). Institute of Plant Protection, Poznan. Poland, 1968-1971

Studies on the Possibility of Using *Bacillus thuringiensis* for the Control of the Indian-meal Moth, *Plodia interpunctella*, or the Mediterranean Flour Moth, *Ephestia kuehniella*. Institute of Plant Protection, Poznan, Poland, 1964-1967

Interaction of Spore-Forming Bacteria and Viruses in Noctuids. Institute of Plant Protection, Poznan, Poland, 1972-1975

Studies on Microbial Insecticides. Pantnagar, India, 1974-1977

Studies on the Crystalliferous Bacteria Infecting Lepidopterous Pests of Pakistan. University of Karachi, Karachi, Pakistan, 1973-1978

The Parasite-Host Relation Using Nematodes of Genera *Neoaplectana* and *Pristionchus*, and the Family Mermithidae Under Laboratory and Field Conditions. Institute of Ecology, Warsaw, Poland, 1971-1976

Virus and Bacterial Diseases Affecting Certain Lepidopterous Agricultural Pests: *Prodenia litura*, *Pyrausta nubilalis*, *Heliothis zea*, *Pectinophora gossypiella*, and *Pieris rapae*. University of Cairo, Giza, Egypt, 1973-1978

Vatican Pontifical Academy of Sciences

Participant, Symposium on the Use of Natural Products to Control Insects, Rome, 1976

World Health Organization (WHO)

Member, Advisory Panel on Insecticides, 1963-1979

Participant, WHO/FAO Meeting on Safety of Insect Viruses, Geneva, 1972

Participant, WHO Conference on Safety of Biological Agents for Arthropod Control, Atlanta, 1973

Honors and Awards

USDA Superior Service Award, 1966

NASA Group Service Award, 1970

III. CONTRIBUTIONS TO THE STUDY OF *BACILLUS THURINGIENSIS*

Arthur Heimpel's contributions to research on *Bacillus thuringiensis* were immensely important. In the early 1950s, Dr. Heimpel became interested in the diseases caused by entomopathogenic bacteria and he began his studies with *B. cereus*, a bacterial species closely related to *B. thuringiensis*. In fact, the disagreement about the species status of *B. thuringiensis* was derived from the proposal that *B. thuringiensis* was more properly classified as a variety of *B. cereus* based on their similarities rather than their differences (Gordon et al., 1973). Indeed, it was not until 1974 that it was listed as a distinct species in Buchanan and Gibbon's 8th editon of *Bergey's Manual of Determinative Bacteriology*. Since Arthur Heimpel's interest in *B. thuringiensis* probably had its genesis during his studies with *B. cereus*, his work with this closely related bacterium could be mentioned here.

One insect of major economic importance in Canada and the United States was the larch sawfly, *Pristiphora erichsonii*. Between 1910 and 1926 most of the larch, *Larix laricina*, was destroyed by a country-wide infestation in Canada. In 1948 a wave of infestation spread from Manitoba east into Ontario (Heimpel, 1954a). Surveys for disease organisms effective against the larch sawfly were made in northwestern Ontario between 1949 and 1952, although, as Heimpel pointed out, the incidence of disease among the feeding stage of the larch sawfly larvae was normally low (Heimpel, 1954a). Bacteria and fungi together usually accounted for about 3 to 5% mortality of the larvae. Most of the bacteria isolated from dead larch sawfly were strains of *B. cereus*. The isolation of *B. cereus* strains pathogenic for the larch sawfly was of considerable interest to Heimpel (1954b), because Steinhaus (1949) had considered strains of *B. cereus* to be pathogens for only some Lepidoptera. In fact, the Heimpel (1954b) publication was the first report of a strain of *B. cereus* (Pr-1017) pathogenic to a hymenopteran. Laboratory and field feeding tests resulted in mortalities as high as 60% and 38%, respectively.

B. cereus was also considered to be the best choice of available pathogens for laboratory experimentation in microbial control of the larch sawfly, because it was relatively successful as a control agent in the field. Thus, Heimpel chose for his doctoral studies the elucidation of the mode of action of *B. cereus* during the process of invasion and destruction of larch sawfly larvae. The results would be used, if possible, to develop a rapid screening test for the most virulent strains of *B. cereus* among isolates obtained in the field. His investigations were based on the assumption that "the mode of action in the insect was of an enzymatic, or at least a biochemical nature." Heimpel (1954a) believed this premise was reasonable; Mattes (1927) first suggested it, and Heimpel surmised that most bacterial pathogens of vertebrates cause damage or death to the host by enzymatic actions or through toxins. It was thought that *B. cereus* and a related species (some strains of *B. anthracis* were the only species in the genus that produce phospholipases (Colmer, 1948; McGaughey and Chu, 1948). Lastly, some strians of *B. cereus* readily killed larch sawfly larvae; those that did not produce phospholipase were avirulent when tested. Heimpel published four papers between 1955-1957 on his extensive studies of *B. cereus* and the phospholipase it produced (Heimpel, 1955a,b, 1956; Kushner and Heimpel, 1957).

The first indication that lecithinase might be important in insect disease was by Toumanoff (1953) who showed a similarity between the lecithinase produced by *B. thuringiensis* var. *alesti* and the α toxin of *Clostridium perfringens*. However, in his work Heimpel (1954a,b; 1955a,b) demonstrated the coincidence between pH optimum (6.6-7.4) for lecithinase and the pH inside the gut of sawfly larvae: species of sawflies whose gut pH was higher than the optimum for the enzyme were not susceptible to *B. cereus*, and histological examination of the insect gut revealed no enzymic destruction; also, species of *Bacillus* that were incapable of producing lecithinase were not found to be pathogenic to the larch sawfly. In addition, Heimpel (1955a) tested *B. cereus* against other insect species and found that resistant species had an alkaline pH in the midgut. Thus, Heimpel concluded that the alkaline condition in the gut of some insects seemed to be a limiting factor in both growth of the organism and lecithinase activity.

In this important work Heimpel (1955a,b) determined the pH in the gut and the blood of 11 species of Hymenoptera and 2 of Lepidoptera. The larvae were examined in their later instars, after ecdysis, after starvation, and as mature larvae. The gut pH was found to change regionally during development, but the blood pH remained relatively unchanged. Heimpel also found that about 3 μg of a lecithinase preparation produced an LD_{50} for fifth-instar larvae of the larch sawfly, further substantiating his explanation of the mode of action. Active enzyme extracted from broth cultures of *B. cereus* was found to be toxic to larch sawfly larvae. The bacterium also could enter the blood through a rupture in the midgut wall after the gut was seriously damaged.

Significantly, Smirnoff and Heimpel (1961a) reported the isolation of a strain of *B. thuringiensis* var. *thuringiensis* from the larch sawfly that caused low mortality. The toxemia was not due to crystal-protein, because the larvae continued to feed for 3 days after ingesting the bacterium. Presumably, virulence was the result of the phospholipase C produced by this strain.

These studies on mode of action conclusively demonstrated that the insect gut was the first line of defense against invading organisms, and Heimpel's studies on reactions in the gut and their effect on the invasion of bacteria resulted in a publication on the pH of the gut contents and blood of more than 100 species of insects (Heimpel, 1961). These initial works with *B. cereus* gave a much needed impetus to the philosophy and need in insect pathology for investigating the many parameters involved with the mode of action of insect pathogens. Art Heimpel and his colleagues subsequently extended this philosophy to include mode of action studies on the crystal-forming bacteria.

As mentioned earlier, *B. thuringiensis* is closely related taxonomically to *B. cereus*, except that the species *B. thuringiensis* is characterized as producing parasporal crystal bodies whereas the "*B. cereus* group" is not (Heimpel, 1976a). From the beginning Art Heimpel was involved in the resolution of taxonomic criteria for the crystalliferous bacteria. After investigations of numerous isolates, he introduced a taxonomic key for this group of bacteria in 1958 and 1960 (Heimpel and Agus, 1958; Heimpel and Angus, 1960a). He later published an improved key (Heimpel, 1967a) that has been widely quoted in the literature. In 1967 Heimpel also proposed a scheme of nomenclature for the various toxins produced by certain serotypes of *B. thuringiensis* (Heimpel, 1967b).

In the 1950s, Art Heimpel and his colleague Tom Angus at the Insect Pathology Research Institute, Sault Ste. Marie, Ontario, Canada, began in earnest to study the site and mode of action of crystalliferous bacteria in lepidopterans (Angus and Heimpel, 1956, 1959; Heimpel and Angus, 1959). Heimpel and Angus (1959) placed susceptible insects into three groups according to their response to *B. thuringiensis*. Those that exhibited general paralysis after ingestion of the *B. thuringiensis* δ-endotoxin and showed a blood pH change were type I insects, e.g., *Bombyx mori*. Type II was representative of most lepidopterans: these insects suffered gut paralysis shortly after feeding on the *B. thuringiensis* δ-endotoxin, but there was no gut leakage and, consequently, no change in blood pH or general paralysis; the insects starved and died from a bacteriaremia. Type III insects succumbed to a normal whole preparation of *B. thuringiensis:* both spores and *B. thuringiensis* δ-endotoxin were necessary to cause death, but large doses of *B. thuringiensis* β-exotoxin (fly toxin) alone could kill. This group was represented by *Anagasta kuehniella* and *Lymantria dispar* (Burgerjon and de Barjac, 1960).

In their work Heimpel and Angus (1959) demonstrated for the first time that the site of action resided with the midgut epithelial cells, where a devastating disruption of the midgut cells could be observed histologically within 60 min of feeding on a parasporal crystal preparation of *B. thuringiensis*. Blood pH changes measured during intoxication could be atributed to leakage of the alkaline gut contents across the disrupted epithelial lining into the hemocoel, thus contributing to the general paralysis and cessation of feeding they noted in Type I insects (Angus and Heimpel, 1956, 1959; Heimpel and Angus, 1959). In their now classical paper, Heimpel and Angus (1959) suggested that the histological changes noted did not result directly from the action of the *B. thuringiensis* δ-endotoxin but rather from a specific substrate of the toxin which existed in the cell cementing substances. The subsequent breakdown of the cells then would be an indirect effect of the toxin and might more directly be caused by the action of the highly alkaline contents of the midgut on disorganized cells. The importance of this pioneering work cannot be overemphasized, because it gave impetus to the many subsequent investigations on the structure and molecular mode of action leading to the disruption of the midgut cell wall by the toxic parasporal crystal protein.

During the 18 years he was Chief of the Insect Pathology Laboratory at Beltsville, Heimpel guided a number of research efforts to further the understanding of the pathological mechanisms of *B. thuringiensis*. In 1964 his laboratory first identified the insect cell cementing substance as a mucopolysaccharide without sulfate having similar chemistry and construction to mammalian hyaluronic acid (Estes and Faust, 1964). Investigations on the chemistry and structure of the *B. thuringiensis* δ-endotoxin and the *B. thuringiensis* β-exotoxin were also in progress under his direction, and the results of those investigations were reported in several scientific journals (Cantwell et al., 1964; Faust and Estes, 1966; Faust et al., 1967, 1971a,b, 1973, 1974b; Faust, 1968; Faust and Dougherty, 1969). These studies also established that in the gut of the lepidopterous insect initial dissolution of the toxic crystals produced by *B. thuringiensis* occurred through action of the inorganic components (e.g., potassium carbonate) and later secondary degradation and release of the toxin was from attack by proteolytic enzyme(s). The work later resulted in the isolation of five toxic components having molecular weights of 2.3×10^5, 6.7×10^4, 3.02×10^4, $\sim 5.0 \times 10^3$, and $\sim 1.0 \times 10^3$. Subsequent research led to a possible explanation for the primary molecular mode of action of the toxin in the susceptible insect. Heimpel and his colleagues also demonstrated that the toxic moieties of the crystal produced by *B. thuringiensis* interacted with certain metabolic proteins and enzymes. They studied the inhibition of trypsin by the toxin, describing the kinetics of the reaction and enabling the development of a chemical method of standardization to predict the LD_{50} for insects where the In_{50} (50% enzyme inhibition) of a toxic preparation was determined. Later, Heimpel directed research aimed at explaining the molecular mode of action of the *B. thuringiensis* δ-endotoxin (Louloudes and Heimpel, 1969; Faust et al., 1974a; Travers et al., 1976). The research culminated in the demonstration that the initial molecular mode of action of *B. thuringiensis* δ-endotoxin might be an uncoupler of oxidative phosphorylation (ATP production), fatally speeding up respiratory action without linking it to energy conservation in the midgut of susceptible insects (Faust et al., 1974a; Travers et al., 1976). Heimpel and his colleagues correlated the biochemical and histological results of this work and that of other researchers to explain the overall pathology of this commercially valuable toxin (Faust et al., 1978). Table 1 summarizes these reports.

Arthur Heimpel also studied the association and identification of crystalliferous bacteria in earthworms. For example, Smirnoff and Hemipel (1961b) found that the earthworm, *Lumbricus terrestrus*, was susceptible to infection by *B. thuringiensis* var. *thuringiensis*. The bacteria were shown to penetrate the gut and enter the body cavity, causing a septicemia. This finding was surprising, because the bacterial organism ahd been tested with negative results against a large variety of vertebrate and invertebrate animals (Heimpel and Angus, 1960b). The results pointed out the need for studies on the survival of *B. thuringiensis* spores in the soil and the stability of the crystal under field conditions. It was stressed, however, that earthworms were not likely to be seriously affected by the amount of *B. thuringiensis* that would be used in control operations against insect pests. Later, Heimpel (1966b) described several specimens of the earthworm *Eisenia foetida* that were afflicted with a "blister disease" containing crystalliferous bacteria in all the lesions. The strains isolated were identified as *B. thuringiensis* var. *thuringiensis*. The association of this strain with the lesions suggested that it may cause the pathological condition. The earthworms in question were reared in pits on rotted cottonseed hulls that contained abrasive materials capable of lacerating the worms. Heimpel suggested that the bacteria might have gained entrance to the worms body via such abrasions, which could account for the local site of infection.

Arthur Heimpel was also involved in the selection of *B. thuringiensis* for studies to determine how spaceflight conditions affects biological systems during the Apollo 16 junar mission (Simmonds et al., 1974). The spores of *B. thuringiensis* var.

TABLE 1. Correlation of the Temporal Sequences of the Biochemical and Histological Effects from *Bacillus thuringiensis* (B.t.) δ-Endotoxin Intoxication in Certain Lepidoptera Larvae.

TIME (min)	BIOCHEMICAL EFFECTS	HISTOLOGICAL EFFECTS
0-1	Degradation of B.t. δ-endotoxin. a/, b/ Glucose uptake stimulated but amino acid; carbonate ion, mono- and divalent cation transport unchanged. c/, d/ Acceleration O_2 uptake in H^+ transport system. e, f/	
1-5	Maximum glucose uptake. c/, d/ Maximum acceleration O_2 uptake, uncoupling of oxidative phosphorylation and cessation of ATP production. e/, f/	
5-10	Cessation of glucose uptake, no changes in K^+, Na^+, Ca^+, Mg^+ in gut tissue. c/, d/	Irregular swelling and distortion of gut microvilli. g/
10-20	General breakdown of permeability control; K^+ and other ions increase in hemolymph; loss of osmotic integrity. d/, h/, i/	Epithelial gut cells swell, protrusions develop, endoplasmic reticulum disrupts; mitochondria increase in size, cristae disintegrate. g/, h/, 1/

(continued)

TABLE 1. (continued)

TIME (min)	BIOCHEMICAL EFFECTS	HISTOLOGICAL EFFECTS
After 20	Paralysis of wilkworm gut. m/, n/, o/	Epithelial cells disrupt, lyse, and slough off into lumen. j/, k/, 1/, p/, q/

a/ Fast and Videnova, 1974.
b/ Fast, 1975.
c/ Fast and Donaghue, 1971.
d/ Fast and morrison, 1972.
e/ Faust et al., 1974a.
f/ Travers et al., 1976.
g/ Ebersold et al., 1977.
h/ Fast and Angus, 1965.
i/ Louloudes and Heimpel, 1969.
j/ Heimpel and Angus, 1959.
k/ Sutter and Raun, 1967.
1/ Angus, 1970.
m/ Hannay, 1953.
n/ Angus, 1954.
o/ Young and Fitz-James, 1959.
p/ Angus and Heimpel, 1959.
q/ Hoopingarner and Materu, 1964.

thuringiensis were flow in the Apollo 16 experiment on Microbial Response to Space Environment and were examined for viability and alterations in the various toxins it produced. In this program, exposure of the spores to full sunlight in space resulted in a significant reduction in viability, which agreed with ground-based studies. However, exposure of the spores to space conditions without light or space conditions in conjunction with solar ultraviolet irradiation at peak wavelengths of 254 and 280 nm did not alter survival rates. Cultures of *B. thuringiensis* returned from spaceflight appeared to be unaffected in their ability to produce the three toxins affecting insects.

In 1978, Dr. Heimpel suggested that the investigations of the Insect Pathology Laboratory should include studies of the genetic mechanisms controlling δ-endotoxin production. These investigations are presently in a very active stage at the laboratory. Several publications on this subject have appeared, particularly in the area of extrachromosomal DNA identification in *B. thuringiensis* strains that may be useful in genetic manipulation of the δ-endotoxin (Faust et al., 1979; Faust, 1979; Faust and Vaughn, 1979; Faust, 1980; Faust and Travers, 1981; Iizuka et al., 1981a,b). The identification of extrachromosomal DNA elements idigenous to entomopathogenic bacteria and development of DNA transformation systems may have far-reaching implications for genetic manipulation, elucidation of pathological processes, and development of improved new entomopathogenic bacteria as biological control agents.

IV. OTHER CONTRIBUTIONS TO THE SCIENCE OF INSECT PATHOLOGY

Arthur Heimpel was involved not only in field trials, which contributed the information necessary for the formulation and registration of *B. thuringiensis* in several countries, but also in the safety testing of insect viruses and in the development of the protocols for the various tests necessary for their registration as microbial insect control agents. The appendix gives a few titles that attest to Heimpel's involvment and interest in this area.

Dr. Heimpel was also involved in the safety testing and assurance of safety to non-target organisms of several other microbials, namely the Japanese beetle pathogen, *Bacillus popilliae* (see Heimpel and Hrubant, 1973; Thompson and Heimpel, 1974), and the mycoacaricide, *Hirsutella thompsonii* (see McCoy and Heimpel, 1980). Additionally, he was a research member of the Lower Animal Test Team, Lunar Receiving Laboratory, NASA, Houston, Texas, that examined the lunar material returned from the first manned landing for the presence of replicating agents possibly harmful to life on earth (Benschoter et al., 1970).

Although he and his colleagues contributed to the understanding of bacterial and viral pathogenesis, including the development of systems for *in vitro* production of polyhedrosis viruses (Goodwin et al., 1970, 1973; Vaughn et al., 1973; Vaughn, 1976; Adams et al., 1977), he was equally concerned with efforts at identifying and studying new pathogens for insects. For example, he described the nuclear polyhedrosis of the cabbage looper, *Trichoplusia ni* (see Heimpel and Adams, 1966). In particular, he discovered a mixed population of nuclear polyhedrosis virus specific for *T. ni*. One type, consisting of single rods embedded in the polyhedron, affected all susceptible tissues of the body but mainly the tissues of the hemocoel. The second type, forming bundles of rods in the polyhedron, was frequently found in the nuclei of the gut cells; it also infected tissues of the hemocoel. The discovery demonstrated that two different "species" of nuclear polyhedrosis viruses co-existed in the same insect. It prompted a later study giving further proof that the virus "orders" the cell to produce polyhedral protein and that the insect cell does not produce it as some form of protection (Tompkins et al., 1969).

Later, Dr. Heimpel was involved in the discovery and description of the pathogenesis of rhabdovirus-like particles in Mexican bean beetles, *Epilachna varivestis*, and house crickets (Adams et al., 1979a,b, 1980). These reports were the first description of a rhabdovirus-like pathogen in coleopteran insects.

More recently, Art Heimpel initiated studies on the mode of action of strains of *Serratia marcescens*, one of the most harmful bacteria infecting insect colonies. He was particularly interested in developing therapeutic methods for eliminating the organism from mass-reared insect colonies. These studies were to lead to the control of this organism in large colonies of insects needed for investigations of the pathogenesis of other entomopathogens. His untimely death prohibited his realization of this goal.

V. CONCLUSIONS

Dr. Arthur Heimpel is recognized as an outstanding pioneer insect microbiologist and a source of information on many aspects of insect pathology, including the application of the many fundamental and applied studies carried out in this field. He also recognized and, for many years, suggested that the goals of insect pathology could be substantially enhanced by the computerization of international sources of literature on insect pathology and microbial control and the establishemnt of international banks of microorganisms used or potentially useful for insect control. The importance of the latter suggestion was made clear in his presidential address given at the VIIth Annual Meeting of the Society for Invertebrate Pathology, held in Tempe, Arizona:

> "I think everyone would agree that one of the primary requirements for good research is the positive identification of the organisms that are to be studied; not only the identification of the organism is important, but the establishment of the type species and strains and the preservation and storage of these organisms for future workers is absolutely essential. In the past, valuable or interesting organisms have been studied, reported, and then lost. This will continue to happen unless something is done to retain and maintain the cultures reported in the literature... Granted there are several organizations in the world that collect and identify and preserve these isolates, however, there are not enough of these to insure that we do not lose valuable specimens."

The impetus of Arthur Heimpel's efforts continue and his impact in this field is still felt. He once stated to me, as he surely did to many others, that "it is exhilarating to realize how far we have advanced in the last quarter century; it is sobering to find how much more we have to accomplish in the future."

VI. APPENDIX

A Bibliography of Research Papers, Scientific Reviews, and Other Articles Authored or Co-authored by Arthur M. Heimpel.

1. MacLeod, D. M. and Heimpel, A. M. (1950). An attempt to establish species of the genus *Beauveria* on a population of the larch sawfly *Pristiphora erichsonii* (Htg.). *Annu. Tech. Rep., Forest Insect Laboratory, Canadian Forestry Service,* Sault Ste. Marie, Ontario.

2. Heimpel, A. M. (1950, 1951, 1952). Investigations of pathogens of the larch sawfly. *Annu. Tech. Rep., Forest Insect Laboratory, Canadian Forestry Service,* Sault Ste. Marie, Ontario.

3. Heimpel, A. M. (1953). Unusual predation of the larch sawfly. *Can. Dep. Agric. For. Biol. Div. Bi-monthly Prog. Report 9,* 2.

4. Heimpel, A. M. (1954). Investigations of the mode of action of strains of *Bacillus cereus*. Doctoral Dissertation. Queen's University, Kingston, Ontario. 154 pp.

5. Heimpel, A. M. (1954). A strain of *Bacillus cereus* Fr. and Fr. pathogenic for the larch sawfly, *Pristiphora erichsonii* (Htg.). *Can. Entomol.*, 86, 73-77.

6. Heimpel, A. M. (1954). An apparatus for mounting and holding insects. *Can. Entomol.* 86, 470.

7. MacLeod, D. M. and Heimpel, A. M. (1955). Fungal and bacterial pathogens of the larch sawfly. *Can. Entomol.* 87, 128-131.

8. Heimpel, A. M. (1955). The pH in the gut and blood of the larch sawfly, *Pristiphora erichsonii* (Htg.) and other insects with reference to the pathogenicity of *Bacillus cereus* Fr. and Fr. *Can. J. Zool.* 33, 99-106.

9. Heimpel, A. M. (1955). Investigations of the mode of action of strains of *Bacillus cereus* Fr. and Fr. pathogenic for the larch sawfly, *Pristiphora erichsonii* (Htg.) *Can. J. Zool.* 33, 311-326.

10. Heimpel, A. M. (1955). Pathogenicity of the bacterium *Serratia marcescens* Bizio for insects. *Can. Dep. Agric. For. Biol. Div., Bi-monthly Prog. Report II.* 3, 1.

11. Heimpel, A. M. (1956). Further observations on the pH in the gut and blood of Canadian forest insects. *Can. J. Zool.* 34, 210-212.

12. Angus, T. A. and Heimpel, A. M. (1956). An effect of *Bacillus sotto* on the larvae of *Bombyx mori*. *Can. Entomol.* 88, 138-139.

13. Kushner, D. J. and Heimpel, A. M. (1957). Lecithinase production by strains of *Bacillus cereus* Fr. and Fr. pathogenic for the larch sawfly, *Pristiphora erichsonii* (Htg.). *Can. J. Microbiol.* 3, 547-551.

14. Heimpel, A. M. and Angus, T. A. (1958). Recent advances in the knowledge of some bacterial pathogens of insects. *Proc. X Int. Congr. Entomol.* (1956), 4, 711-722.

15. Heimpel, A. M. and Angus, T. A. (1958). The taxonomy of insect pathogens related to *Bacillus cereus* Fr. and Fr. *Can. J. Microbiol.* 4, 531-541.

16. Heimpel, A. M. (1958). Notes on methods for rearing two Canadian forest insects. *Ann. Rep. Entomol. Soc. Ont.* 88, 42-43.

17. Angus, T. A. and Heimpel, A. M. (1958). Further observations of the action of *Bacillus sotto* toxin. *Can. Dept. Agric. For. Biol. Div., Bi-monthly Prog. Rep.* 14, 1-2.

18. Angus, T. A. and Heimpel, A. M. (1959). Inhibition of feeding and blood pH changes in lepidopterous larvae infected with crystal-forming bacteria. *Can. Entomol.* 91, 352-358.

19. Heimpel, A. M. and West, A. S. (1959). Notes on the pathogenicity of *Serratia marcescens* Bizio for the cockroach *Blatella germanica* L. *Can. J. Zool.* 37, 169-172.

20. Heimpel, A. M. and Angus, T. A. (1959). The site of action crystalliferous bacteria in Lepidoptera larvae. *J. Insect Pathol.* 1, 152-170.

21. Angus, T. A. and Heimpel, A. M. (1959). Microbial insecticides. *Can. Dep. Agric. Res. Farmers.* 4 (2), 12-13.

22. Angus, T. A. and Heimpel, A. M. (1960). The bacteriological control of insects. *Proc. Entomol. Soc. Ont.* 90, 13-21.

23. Heimpel, A. M. and Angus, T. A. (1960). On the taxonomy of certain entomogenous crystalliferous bacteria. *J. Insect Pathol.* 2, 311-319.

24. Heimpel, A. M. and Angus, T. A. (1960). Bacterial insecticides. *Bacteriol. Rev.* 24, 266-288.

25. Heimpel, A. M. (1961). Pathogenicity of *Bacillus cereus* Fr. and Fr. and *Bacillus thuringiensis* Berliner varieties for several species of sawfly larvae. *J. Insect Pathol.* 3, 271-273.

26. Heimpel, A. M. (1961). The application of pH determinations to insect pathology. *Proc. Entomol. Soc. Ont.* 91, 52-76.

27. Smirnoff, W. and Heimpel, A. M. (1961). A strain of *Bacillus thuringiensis* isolated from the larch sawfly. *Pristiphora erichsonii* (Htg.). *J. Insect Pathol.* 3, 347-351.

28. Smirnoff, W. and Heimpel, A. M. (1961). Notes on the pathogenicity of *Bacillus thuringiensis* var. *thuringiensis* Berliner for the earthworm, *Lumbricus terrestria* Linnaeus. *J. Insect Pathol.* 3, 403-408.

29. Kinghorn, J. M., Fisher, R. A., Angus, T. A., and Heimpel, A. M. (1961). Tests of a microbial insecticide against forest defoliators. *Can. Dep. Agric. For. Biol. Div., Bi-monthly Prog. Rep.* 17, 1-4.

30. Robertson, S. W., Jr. and Heimpel, A. M. (1962). Crystal preparations from commercial *Bacillus thuringiensis* var. *sotto* Concentrates. *J. Insect Pathol.* 4, 273-274.

31. Heimpel, A. M. (1962). Microbial insecticides. *Proc. North Cent. Branch Entomol. Soc. Am.* 17, 80-84.

32. Heimpel, A. M. and Angus, T. A. (1963). Diseases caused by certain spore-forming bacteria. *In* "Insect Pathology: An Advanced Treatise" (E. A. Steinhaus ed.) 2, 21-73. Academic Press, New York.

33. Heimpel, A. M. (1964). General aspects of bacteriological control. *Entomophaga Memoire* 2, 23-33.

34. Heimpel, A. M. (1963). The status of *Bacillus thuringiensis*. *Advances in Chemistry Sciences*. Pesticide Control and Eradication Symp., Am. Chem. Soc., Atlantic City. (1962). 41, 64-74.

35. Heimpel, A. M. (1963). Introductory remarks on microbial control. *Developments in Industrial Microbiology*. AIBS, Washington, D. C. 4, 131-136.

36. Cantwell, G. E., Heimpel, A. M., and Thompson, M. J. (1964). The production of an exotoxin by various crystal-forming telated to *Bacillus thuringiensis* var. *thuringiensis*. *J. Insect Pathol.* 6, 466-485.

37. Heimpel, A. M. (1965). The specificity of the pathogen *Bacillus thuringiensis* var. *thuringiensis* for insects. *Proc. XII Int. Congr. Entomol.*, London, England, 1964, p. 736.

38. Heimpel, A. M. and Harshbarger, J. C. (1965). Symposium on microbial control. V. Immunity in insects. *Bacteriological Reviews* 29, 397-405.

39. Heimpel, A. M. (1965). Microbial control of insects. *World Pest Rev.* 4, 150-161.

40. Ignoffo, C. M. and Heimpel, A. M. (1965). The nuclear polyhedrosis virus of *heliothis zea* (Boddie) and *Heliothis virescens* (Fabricius). V. Toxicity-pathogenicity of virus to white mice and guinea pigs. *J. Invertebr. Pathol.* 7, 329-340.

41. Heimpel, A. M. (1966). Insect pathology, present and future. Symposium on alternate methods of insect control (E. F. Knipling, ed.), *U.S. Dept. Agric., Agric. Res. Serv.*, Publ. 33-110. Washington, D.C. p. 70-75.

42. Heimpel, A. M. (1966). Exposure of white mice and guinea pigs to the nuclear-polyhedrosis virus of the cabbage looper, *Trichoplusia ni*. *J. Invertebr. Pathol.* 8, 98-102.

43. Heimpel, A. M. and Adams, J. R. (1966). A new nuclear polyhedrosis of the cabbage looper, *Trichoplusia ni*. *J. Invertebr. Pathol.*, 8, 340-346.

44. Heimpel, A. M. (1966). A crystalliferous bacterium associated with a "Blister Disease" in the earthworm *Eisenia foetida* (Savigny). *J. Invertebr. Pathol.* 8, 295-298.

45. Heimpel, A. M. (1967). A proposed society for invertebrate pathology (Addendum). *J. Invertebr. Pathol.*, 9: iv.

46. Heimpel, A. M. (1967). The problem associated with the standardization of insect viruses. *Proc. Colloq. Insect Pathol.*, Wageningen, Holland, Sept. 1966, p. 355-359.

47. Heimpel, A. M. (1967). A critical review of *Bacillus thuringiensis* var. *thuringiensis* and other crystalliferous bacteria. *Annu. Rev. Entomol.* 12, 287-322.

48. Heimpel, A. M. and Buchanan, L. K. (1967). Human feeding tests using the nuclear polyhedrosis virus of *Heliothis zea* (Boddie). *J. Invertebr. Pathol.* 9, 55-57.

49. Heimpel, A. M. (1967). A taxonomic key proposed for the species of the crystalliferous bacteria. *J. Invertebr. Pathol.* 9, 364-375.

50. Heimpel, A. M. (1967). Insektenvirus-Präparate zur mikrobiologischen Schädlingsbekämpfung. *Umsch. Wiss. Tech.* 23, 759-763.

51. Heimpel, A. M. (1967). Progress in developing insect viruses as microbial control agents. *Proc. Joint U.S.-Jpn. Semin. Microbiol. Control Insect Pests*, Fukuoka, Japan. p. 51-61.

52. Faust, R. M., Adams, J. R., and Heimpel, A. M. (1967). Dissolution of the toxic parasporal crystals from *Bacillus thuringiensis* var. *pacificus* by gut secretions of the silkworm *Bombyx mori*. *J. Invertebr. Pathol.* 9, 488-499.

53. Harshbarger, J. C. and Heimpel, A. M. (1968). Effect of Zymosan on phagocytosis in larvae of the greater wax moth, *Galleria mellonella*. *J. Invertebr. Pathol.* 10, 176-179.

54. Heimpel, A. M. (1968). Letter of reply to Martin H. Rogoff on "Usefulness of a newly proposed key for species of crystalliferous bacteria." *J. Invertebr. Pathol.* 10, 454-456.

55. Louloudes, S. J. and Heimpel, A. M. (1969). Mode of action of *Bacillus thuringiensis* toxic crystals in larvae of the silkworm, *Bombyx mori*. *J. Invertebr. Pathol.* 14, 375-381.

56. Tompkins, G. J., Adams, J. R., and Heimpel, A. M. (1969). Cross infection studies with *Heliothis zea* using nuclear-polyhedrosis viruses from *Trichoplusia ni*. *J. Invertebr. Pathol.* 14, 343-357.

57. Wells, F. E. and Heimpel, A. M. (1970). Replication of insect viruses in bacterial hosts. *J. Invertebr. Pathol.* 16, 301-304.

58. Benschoter, C. A., Allison, T. C., Boyd, J. F., Brooks, M. A., Campbell, J. W., Groves, R. O., Heimpel, A. M., Mills, H. E., Ray, S. M., Warren, J. W., Wolf, K. E., Wood, E. M., Wrenn, R. T., and Zein-Eldin, Z. (1970). Apollo 11: Exposure of lower animals to lunar material. *Science* 169, 470-472.

59. Heimpel, A. M. (1971). Safety of insect pathogens for man and vertebrates. *In* "Microbial Control of Insects and Mites" (D. Burges & N. Hussey, eds.), p. 469-489. Academic Press, London, England.

60. Faust, R. M., Dougherty, E. M., Heimpel, A. M., and Reichelderfer, C. F. (1971). Standardization of the δ-endotoxin produced by several varieties of *Bacillus thuringiensis*. I. Enzyme kinetics of the trypsin-azoalbumin-δ-endotoxin system. *J. Econ. Entomol.* 64, 610-615.

61. Faust, R. M., Dougherty, E. M., Heimpel, A. M., and Reichelderfer, C. F. (1971). Standardization of the δ-endotoxin produced by several varieties of *Bacillus thuringiensis*. II. Enzyme inhibition (In_{50} as a method of standardizing commerical preparations of the δ-endotoxin. *J. Econ. Entomol.* 64, 615-621.

62. Heimpel, A. M. (1972). Safety of insect pathogens. *Proc. WHO-FAO Meeting on Safety of Mocrobial Control Agents*, Geneva, Switzerland. p. 628-639.

63. Heimpel, A. M. (1972). The development and progress of insect microbial control. *Proc. Inst. Biol. Control*, Mississippi State Univ. p. 528-531.

64. Thomas, E. D., Reichelderfer, C. F., and Heimpel, A. M. (1972). Accumulation and persistence of a nuclear polyhedrosis virus of the cabbage looper in the field. *J. Invertebr. Pathol.* 20, 157-164.

65. Taylor, G. R., et al. (1972). Microbial response to space environment *In* "Apollo 16 Preliminary Science Report", NASA Spec. Publi. 315, U.S. Govt. Printing Off. Washington, D. C. p. (27-11)-(27-17).

66. Lightner, D. V., Proctor, R. R., Sparks, A. K., Adams, J. R., and Heimpel, A. M. (1973). Testing penaeid shrimp for susceptibility to an insect nuclear polyhedrosis virus. *Environ. Entomol.* 2, 611-613.

67. Heimpel, A. M. and Hrubant, G. G. (1973). Medical examination of humans exposed to *Bacillus popilliae* and *Popillia japonica* during production of commercial milky disease spore dust. *Environ. Entomol.* 2, 793-795.

68. Adams, J. R., Faust, R. M., and Heimpel, A. M. (1973). X-ray analysis of insect viruses and crystals of *Bacillus thuringiensis*. *Proc. V Int. Colloq. Insect Pathol. Microbiol. Control.* Oxford, England. p. 33.

69. Thomas, E. D., Reichelderfer, C. F., and Heimpel, A. M. (1973). The effect of soil pH on the persistence of cabbage looper nuclear polyhedrosis virus in soil. *J. Invertebr. Patho.* 21, 21-25.

70. Heimpek, A. M., Thomas, E. D., Adams, J. R., and Smith, L. J. (1973). The presence of nuclear polyhedrosis viruses of *Trichoplusia ni* on cabbage from the market shelf. *Environ. Entomol.* 2, 72-75.

71. Thomas, E. D., Heimpel, A. M., and Adams, J. R. (1974). Determination of the active nuclear polyhedrosis virus content of untreated cabbages. *Environ. Entomol.* 3, 980-910.

72. Simmonds, R. C., Wrenn, R. J., Heimpel, A. M., and Taylor, G. R. (1974). Postflight analysis of *Bacillus thuringiensis* organisms exposed to space flight conditions on Apollo 16. *Aerosp. Med.* November, 1244-1247.

73. Thompson, J. V. and Heimpel, A. M. (1974). Microbiological examination of *Bacillus popilliae* product called DOOM. *Environ. Entomol.* 3, 182-183.

74. Heimpel, A. M. (1975). Presidential Address. *J. Invertebr. Pathol.* 25, 1-4.

75. Heimpel, A. M. (1977). Practical applications of insect viruses. *Beltsville Symp. Agric. Res. I. Virology in Agric.* Beltsville, Md., May 1976. Alanheld, Osman and Montclair, New York. p. 101-107.

76. Heimpel, A. M. (1977). The use of viruses in plant protection. *Proc. Symp. Use of Natural Products to Control Insects* (1976). The Vatican Pontif. Acad. of Science, The Vatican, Rome, Italy, 41, 1-23.

77. Faust, R. M., Travers, R. S., and Heimpel, A. M. (1978). Correlation of the biochemical and histological temporal sequence in *Bacillus thuringiensis* δ-endotoxin intoxication. *Prog. Invertebr. Pathol.* (J. Weiser, ed.), Agricultural College Campus Press, Prague, Czechoslovakia. p. 63-64.

78. Adams, J. R., Tompkins, G. J., Heimpel, A. M., and Dougherty, E. M. (1978). Electron investigations on rhabdovirus-like particles in Mexican bean beetles and crickets. *Proc. IX Int. Congr. Electron Microsc.* Toronto, 1978, 2, 378-379.

79. Adams, J. R., Dougherty, E. M., and Heimpel, A. M. (1978). Rhabdovirus-like particles isolated from crickets: Isolation, pathogenesis and characterization. *Abst. IV Int. Congr. Virology,* The Hague, The Netherlands, August 30 - September 6, 1978. p. 573.

80. Hostetter, D. L., Biever, K. D., Heimpel, A. M., and Ignoffo, C. M. (1979). Efficacy of the nuclear polyhedrosis virus of the alfalfa looper against cabbage looper larvae on cabbage in Missouri. *J. Econ. Entomol.* 72, 371-373.

81. Adams, J. R., Dougherty, E. M., and Heimpel, A. M. (1980). Pathogenesis of rhabdovirus-like particles in the house cricket, *Acheta domesticus*. *J. Invertebr. Pathol.* 35, 314-317.

82. McCoy, C. W. and Heimpel, A. M. (1980). Safety of the potential mycoacaricide, *Hirsutella thompsonii*, to vertebrates. *Environ. Entomol.* 9, 47-49.

VII. REFERENCES

Adams, J. R., Goodwin, R. H., and Wilcox, T. A. (1977). Electron microscopic investigations on invasion and replication of insect baculoviruses *in vivo* and *in vitro*. *Biol. Cellulaire.* 28, 261-268.

Adams, J. R., Dougherty, E. M., and Heimpel, A. M. (1978a). Rhabdovirus-like particles isolated from crickets: Isolation, pathogenesis and characterization. *Abst. IV Int. Congr. Virology*, The Hague, The Netherlands, August 30 - September 6, 1978. p. 573.

Adams, J. R., Tompkins, G. J., Heimpel, A. M., and Dougherty, E. (1978b). Electron investigations on rhabdovirus-like particles in Mexican bean beetles and crickets. *In* Proc. IX Int. Congr. Electron Microsc., Toronto. 2, 378-379.

Adams, J. R., Dougherty, E. M., and Heimpel, A. M. (1980). Pathogenesis of rhabdovirus-like particles in the house cricket, *Acheta domesticus*. *J. Invertebr. Pathol.* 35, 314-317.

Angus, T. A. (1954). A bacterial toxin paralyzing silkworm larvae. *Nature.* 173, 545.

Angus, T. A. (1970). Implications of some recent studies of *Bacillus thuringiensis* - A personal preview. IV Int. Colloq. Insect Pathol. Proc. College Park, MD. August 25-28, 1970. pp. 183-190.

Angus, T. A. and Heimpel, A. M. (1956). An effect of *Bacillus sotto* on the larvae of *Bombyx mori*. *Can. Entomol.* 88, 138-139.

Angus, T. A. and Heimpel, A. M. (1959). Inhibition of feeding and blood pH changes in lepidopterous larvae infected with crystal-forming bacteria. *Can. Entomol.* 91, 352-358.

Benschoter, C. A., Allison, T. C., Boyd, J. F., Brooks, M. A., Campbell, J. W., Groves, R. O., Heimpel, A. M., Mills, H. E., Ray, S. M., Warren, J. W., Wolf, K. E., Wood, E. M., Wrenn, R. T., and Zein-Eldin, Z. (1970). Apollo 11: Exposure of lower animals to lunar material. *Science* 169, 470-472.

Buchanan, R. E. and Gibbons, N. E. (1974). Endospore-forming rods and cocci. *In* "Bergey's Manual of Determinative Bacteriology" (R. E. Buchanan and N. E. Gibbons, eds.). pp. 529-575. The Williams and Wilkins Company, Baltimore, MD.

Burgerjon, A. (1965). Le titrage biologique des cristaux de *Bacillus thuringiensis* Berliner par reduction de consommation au laboratoire de la miniere. *Entomophaga* 10, 21-26.

Burgerjon, A. and de Barjac, H. (1960). Nouvelles donnees sur le role de la toxine soluble thermostable produite par *Bacillus thuringiensis* Berliner. *C. R. Acad. Sci.* Paris. 251, 911-912.

Cantwell, G. E., Heimpel, A. M., and Thompson, J. J. (1964). The production of an exotoxin by various crystal-forming bacteria related to *Bacillus thuringiensis* var. *thuringiensis*. *J. Insect Pathol.* 6, 466-485.

Colmer, A. R. (1948). The action of *Bacillus cereus* and related species on the lecithin complex of egg yolk. *J. Bacteriol.* 55, 777-785.

Ebersold, H. R., Luthy, P., and Mueller, M. (1977). Changes in the fine structure of the gut epithelium of *Pieris brassicae* induced by the δ-endotoxin of *Bacillus thuringiensis*. *Bull. Soc. Entomol. Suisse.* 50, 269-276.

Estes, Z. E. and Faust, R. M. (1964). Studies on the mucopolysaccharides of the greater wax moth, *Galleria mellonella*. *Comp. Biochem. Physiol.* 13, 443-452.

Fast, P. G. (1975). Purification of fragments of *Bacillis thuringiensis* δ-endotoxin from hemolymph of spruce budworm. *Bi-monthly Res. Notes Can. For. Serv.* 31, 1-2.

Fast, P. G. and Angus, T. A. (1965). Effects of parasporal inclusions of *Bacillus thuringiensis* var. *sotto* Ishiwata on the permeability of the gut wall of *Bombyx mori* (Linnaeus) larvae. *J. Invertebr. Pathol.* 7, 29-32.

Fast, P. and Donaghue, T. (1971). The δ-endotoxin of *Bacillus thuringiensis*. II. On the mode of action. *J. Invertebr. pathol.* 18, 135-138.

Fast, P. and Morrison, I. (1972). The endotoxin of *Bacillus thuringiensis*. IV. The effect of δ-endotoxin on ion regulation by midgut tissue of *Bombyx mori* larvae. *J. Invertebr. Pathol.*, 20, 208-211.

Fast, P. and Videnova, E. (1974). The δ-endotoxin of *Bacillus thuringiensis*. V. On the occurrence of endotoxin fragments in hemolymph. *J. Invertebr. Pathol.* 18, 135-138.

Faust, R. M. (1968). In vitro chemical reaction of the δ-endotoxin produced by *Bacillus thuringiensis* var. *dendrolimus* with other porteins. *J. Invertebr. Pathol.* 11, 465-475.

Faust, R. M. (1979). Pathogenic bacteria of pest insects as host-vector systems. *Recombinant DNA Technical Bulletin.* DHEW Pub. NIH 79-99. 2, 1-8.

Faust, R. M. (1980). Genetic factors for improving entomopathogenic bacteria: Current research. *Abst. XVI Int. Congr. Entomol.* Kyoto, Japan, p. 189.

Faust, R. M. and Estes, Z. E. (1966). Silicon content of the parasporal crystal of several crystalliferous bacteria. *J. Invertebr. Pathol.* 8, 141-144.

Faust, R. M., Adams, J. R., and Heimpel, A. M. (1967). Dissolution of the toxic parasporal crystals from *Bacillus thuringiensis* var. *pacificus* by the gut secretions of the silkworm, *Bombyx mori*. *J. Invertebr. Pathol.* 9, 488-499.

Faust, R. M. and Dougherty, E. M. (1969). Effects of the δ-endotoxin produced by *Bacillus thuringiensis* var. *dendrolimus* on the hemolymph of the silkworm, *Bombyx mori*. *J. Invertebr. Pathol.* 13, 155-157.

Faust, R. M., Dougherty, E. M., Heimpel, A. M., and Reichelderfer, C. F. (1971a). Standardization of the δ-endotoxin produced by several varieties of *Bacillus thuringiensis*. I. Enzyme kinetics of the trypsin-azoalbumin-δ-endotoxin system. *J. Econ. Entomol.* 64, 610-615.

Faust, R. M., Dougherty, E. M., Heimpel, A. M., Reichelderfer, C. F. (1971b). Standardization of the δ-endotoxin produced by several varieties of *Bacillus thuringiensis*. II. Enzyme inhibition (In_{50}) as a method of standardizing commercial preparations of the δ-endotoxin. *J. Econ. Entomol.* 64, 615-621.

Faust, R. M., Hallam, G. M., and Travers, R. S. (1973). Spectrographic elemental analysis of the parasporal crystals produced by *Bacillus thuringiensis* var. *dendrolimus* and polyhedral inclusion bodies of the fall armyworm, *Spodoptera frugiperda*, nucleopolyhedrosis virus. *J. Invertebr. Pathol.* 22, 478-480.

Faust, R. M., Travers, R. S., and Hallam, G. M. (1974a). Preliminary investigations on the molecular mode of action of the δ-endotoxin by *Bacillus thuringiensis* var. *alesti*. *J. Invertebr. Pathol.* 23, 259-261.

Faust, R. M., Hallam, G. M., and Travers, R. S. (1974b). Degradation of the parasporal crystal produced by *Bacillus thuringiensis* var. *kurstaki*. *J. Invertebr. Pathol.* 24, 365-373.

Faust, R. M., Travers, R. S., and Heimpel, A. M. (1978). Correlation of the biochemical and histological temporal sequence in *Bacillus thuringiensis* δ-endotoxin intoxication. *Prog. Invertebr. Pathol.* (J. Weiser, ed.), Agricultural College Campus Press, Prague, Czechoslovakia. p. 63-64.

Faust, R. M., Spizizen, J., Gage, V., and Travers, R. S. (1979). Extrachromosomal DNA in *Bacillus thuringiensis* var. *kurstaki*, var. *finitimus*, var. *sotto*, and *B. popilliae*. *J. Invertebr. Pathol.* 33, 233-238.

Faust, R. M. and Vaughn, J. L. (1979). Use of tissue culture and genetic engineering technology for development of microbial insect control agents. *Proc. 36th South. Past. Crop Improve. Conf.*, May 1-3, 1979, Beltsville, MD. AR, SEA, USDA p. 8-20.

Faust, R. M. and Travers, R. S. (1981). Occurrence of resistance to neomycin and kanamycin in *Bacillus popilliae* and certain serotypes of *Bacillus thuringiensis*: Mutation potential in sensitive strains. *J. Invertebr. Pathol.* 37, 113-116.

Goodwin, R. H., Vaughn, J. L., Adams, J. R., and Louloudes, S. J. (1970). Replication of a nuclear polyhedrosis virus in an established insect cell line. *J. Invertebr. Pathol.* 16, 284-288.

Goodwin, R. H., Vaughn, J. L., Adams, J. R., and Louloudes, S. J. (1973). The effect of insect cell lines and tissue culture media on polyhedra production. *Misc. Publ. Entomol. Soc. Am.* 9, 66-72.

Gordon, R. E., Haynes, W. C., Pang, C. H-N. (1973). The Genus *Bacillus*. U.S. Dep. Agric., *Agriculture Handb. #427*, Washington, D.C. 283 pp.

Hannay, C. L. (1953). Crystalline inclusions in aerobic spore-forming bacteria. *Nature Lond.* 172, 1004.

Heimpel, A. M. (1954a). Investigations of the mode of action of strains of *Bacillus cereus*. Doctoral Dissertation. Queen's University, Kingston, Ontario. 154 pp.

Heimpel, A. M. (1954b). A strain of *Bacillus cereus* Fr. and Fr. pathogenic for the larch sawfly, *Pristiphora erichsonni* (Htg.) *Can. Entomol.*, 86, 73-77.

Heimpel, A. M. (1955a). The pH in the gut and blood of the larch sawfly, *Pristiphora erichsonii* (Htg.) and other insects with reference to the pathogenicity of *Bacillus cereus* Fr. and Fr. *Can. J. Zool.* 33, 99-106.

Heimpel, A. M. (1955b). Investigations of the mode of action of strains of *Bacillus cereus* Fr. and Fr. pathogenic for the larch sawfly, *Pristiphora erichsonii* (Htg.). *Can. J. Zool.* 33, 311-326.

Heimpel, A. M. (1956). Further observations on the pH in the gut and blood of Canadian forest insects. *Can. J. Zool.* 34, 210-212.

Heimpel, A. M. (1961). The application of pH determinations to insect pathology. *Proc. Entomol. Soc. Ont.* 91, 52-76.

Heimpel, A. M. (1966b). A crystalliferous bacterium associated with a "Blister Disease" in the earthworm *Eisenia foetida* (Savigny). *J. Invertebr. Pathol.* 8, 295-298.

Heimpel, A. M. (1967a). A taxonomic key proposed for the species of the crystalliferous bacteria. *J. Invertebr. Pathol.* 9, 364-375.

Heimpel, A. M. (1967b). A critical review of *Bacillus thuringiensis* var. *thuringiensis* and other crystalliferous bacteria. *Ann. Rev. Entomol.* 12, 287-322.

Heimpel, A. M. and Angus, T. A. (1958). The taxonomy of insect pathogens related to *Bacillus cereus* Frankland and Frankland. *Can. J. Microbiol.* 4, 531-541.

Heimpel, A. M. and Angus, T. A. (1959). The site of action of crystalliferous bacteria in Lepidoptera larvae. *J. Insect Pathol.* 1, 152-170.

Heimpel, A. M. and Angus, T. A. (1960a). On the taxonomy of certain entomogenous crystalliferous bacteria. *J. Insect Pathol.* 2, 311-319.

Heimpel, A. M. and Angus, T. A. (1960b). Bacterial insecticides. *Bacteriol. Rev.* 24, 266-288.

Heimpel, A. M. and Angus, T. A. (1963). Diseases caused by certain spore-forming bacteria. *In* "Insect Pathology: An Advanced Treatise," (E. A. Steinhaus, ed.), 2, 21-73. Academic Press, New York.

Heimpel, A. M. and Adams, J. R. (1966). A new nuclear polyhedrosis of the cabbage looper, *Trichoplusia ni*. *J. Invertebr. Pathol.* 8, 340-346.

Heimpel, A. M. and Hrubant, G. G. (1973). Medical examination of humans exposed to *Bacillus popilliae* and *Popillia japonica* during production of commercial milky disease spore dust. *Environ. Entomol.* 2, 793-795.

Hoopingarner, R. and Materu, M. E. (1964). Toxicology and histopathology of *Bacillus thuringiensis* Berliner in *Galleria mellonella* L. *J. Insect Pathol.* 6, 26-39.

Iizuka, T., Faust, R. M., and Travers, R. S. (1981a). Comparative profiles of plasmid DNA in single and multiple crystalliferous strains of *Bacillus thuringiensis* variety *kurstaki*. *J. fac. Agric. Hokkaido Univ.*, Sapporo, Japan 60, 143-151.

Iizuka, T., Faust, R. M., and Travers, R. S. (1981b). Isolation and partial characterization of extrachromosomal DNA from serotypes of *Bacillus thuringiensis* pathogenic to lepidoptera and diptera larvae by agarose gel electrophoresis. *J. Sericult. Sci. Jpn.* 50, 1-14.

Kushner, D. J. and Heimpel, A. M. (1957). Lecithinase production by strains of *Bacillus cereus* Fr. and Fr. pathogenic for the larch sawfly, *Pristiphora erichsonii* (Htg.). *Can. J. Microbiol.* 3, 547-551.

Louloudes, S. J. and Heimpel, A. M. (1969). Mode of action of *Bacillus thuringiensis* toxic crystals in larvae of the silkworm, *Bombyx mori*. *J. Invertebr. Pathol.* 14, 375-381.

Mattes, O. (1927). Parasitore Kronkheiten der Mehlmottenlarven und Versucke uber ihre verwenborkeit als biologisches Bekempfungsmittel. *Sitzungsber. Ges. Befoerd. Gesamten Naturwiss. Marburg.* 62, 381-417.

McCoy, C. W. and Heimpel, A. M. (1980). Safety of the potential mycoacaricide, *Hirsutella thompsonii*, to vertebrates. *Environ. Entomol.* 9, 47-49.

McGaughey, C. A. and Chu, H. P. (1948). The egg yolk reaction of aerobic sporing bacilli. *J. Gen. Microbiol.* 2, 334-340.

Simmonds, R. C., Wrenn, R. J., Heimpel, A. M., and Taylor, G. R. (1974). Postflight analysis of *Bacillus thuringiensis* organisms exposed to space flight conditions on Apollo 16. *Aerosp. Med.*, November, 1244-1247.

Smirnoff, W. and Heimpel, A. M. (1961a). A strain of *Bacillus thuringiensis* isolated from the larch sawfly, *Pristiphora erichsonii* (Htg.). *J. Insect Pathol.* 3, 347-351.

Smirnoff, W. and Heimpel, A. M. (1961b). Notes on the pathogenicity of *Bacillus thuringiensis* var. *thuringiensis* Berliner for the earthworm, *Lumbricus terrestrus* Linnaeus. *J. Insect Pathol.* 3, 403-408.

Steinhaus, E. A. (1949). *In* "Principles of Insect Pathology." 757 pp. McGraw-Hill Book Co., Inc. New York.

Sutter, G. R. and Raun, E. S. (1967). Histopathology of European corn borer larvae treated with *Bacillus thuringiensis*. *J. Invertebr. Pathol.* 9, 90-103.

Thompson, J. V. and Heimpel, A. M. (1974). Microbiological examination of *Bacillus popilliae* product called DOOM. *Environ. Entomol.* 3, 182-183.

Tompkins, G. J., Adams, J. R., and Heimpel, A. M. (1969). Cross infection studies with *Heliothis zea* using nuclear-polyhedrosis viruses from *Trichoplusia ni*. *J. Invertebr. Pathol.* 14, 343-357.

Toumanoff, C. (1953). Description de quelques souches entomophytes de *Bacillus cereus* Frank. et Frank., avec remarques sur leur action et celle d'autres bacilles dur le jaune d'oeuf. *Ann. Inst. Pasteur.* Paris 85, 90-98.

Travers, R. S., Faust, R. M., and Reichelderfer, C. F. (1976). Effects of *Bacillus thuringiensis* var. *kurstaki* δ-endotoxin on isolated lepidopteran mitochondria. *J. Invertebr. Pathol.* 28, 351-356.

Vaughn, J. L. (1976). The production of nuclear polyhedrosis viruses in large - volume cell cutures. *J. Invertebr. Pathol.* 28, 233-237.

Vaughn, J. L., Adams, J. R., and Wilcox, T. (1973). Infection and replication of insect viruses in tissue culture. *Monogr. Virol.* 6, 25-35.

Young, I. E. and Fitz-James, P. C. (1959). Chemical and morphological studies of bacterial spore formation. II. Spore and parasporal protein formation in *Bacillus cereus* var. *alesti*. *J. Biophys. Biochem. Cytol.* 6, 483-498.

INSPIRATION, SWEAT, AND SERENDIPITY: THE PROOF OF *Bacillus thuringiensis* IN BIOLOGICAL CONTROL

Thomas A. Angus

Canadian Forestry Service
Sault Ste Marie, Ontario
Canada

I. INTRODUCTION

When the Society for Invertebrate Pathology met in College Park, Maryland, for its 1970 annual meeting and for the Fourth International Colloquium on Insect Pathology, the Edward A. Steinhaus memorial address was given by Dr. Edward F. Knipling of the U.S. Department of Agriculture. As he pointed out, we owe a great debt to foresighted individuals such as Edward Steinhaus and Arthur Heimpel, not only for their personal scientific contributions but also for their support to the field of microbial control of insects (Knipling, 1970). It is entirely appropriate that we should honor their contributions now.

It was with a real sense of pride that I received this invitation to present my thoughts on the importance of *Bacillus thuringiensis* in biological control. Ten years ago in a similar invitational paper I indicated that I was about to begin a new career as a Research Manager in the Public Service of Canada (Angus, 1970). I am very much aware that in some quarters we are regarded as renegades and that we are referred to as the enemy, the eunichs, etc. May I plead that we are a necessary evil. Taking on the role of combined den-mother, father-confessor, sheep-dog, amanuensis, judas-goat, and tail-end Charlie for a lovable gang of talented and proud scientific primae donnae leaves little time for personal bench research. Thus, in this presentation there will be much review and little fresh material.

With a light heart, I unprotestingly accepted the somewhat flamboyant title put forward by the Program Committee of the Society for Invertebrate Pathology and only after commencing to prepare the text did I realize the difficulties it posed. I tried then to determine the boundaries of my assignment by a process of elimination, the results of which are presented here.

II. DEVELOPMENTAL HISTORY OF *Bacillus thuringiensis* - A PERSONAL REVIEW

The best known early studies in insect pathology involved principally the fungi and then the protozoans. The identification of viruses as agents of disease in insects did not occur until the late 1890s. One of the first known bacterial pathogen of insects, *Bacillus alvei*, the causative agent of foulbrood of honey bees was described by Cheshire (1885). In this connection, it is tempting to assume that Louis Pasteur (1870), while investigating the pebrine disease of silkworms must have seen larvae killed by strains of *Bacillus thuringiensis*. Indeed, Pasteur (1870) found that dust collected from menageries where silkworm had died of flacherie was capable of bringing on the malady in healthy larvae within a few hours. This point is tantalizing because Pasteur was an acute and gifted observer and yet this observation, so obvious to us now in its implications, did not lead to any endeavors of scientific benefit. However, one must remember that Pasteur was attempting to solve the riddle of pebrine, a highly infectious protozoan disease of silkworm, and he was seeking above all a cure for this malady.

Steinhaus (1956) has noted that the principal chronological steps in the evolution of the idea of microbial control involved realization that (1) insects, such as the silkworm, were subject to disease; (2) the diseases were of a contagious nature; (3) natural outbreaks of disease occurred; (4) insects served as natural hosts for certain species of pathogens; (5) the agents of the diseases could be transmitted from diseased to healthy individuals; (6) the diseases were caused by living agents (microorganisms) that could be dispersed; and (7) these disease-causing microorganisms could be grown in numbers and quantities large enough for field distribution. Some steps of this evolutionary process were pertinent to the development of *Bacillus thuringiensis* as a biocontrol agent and some were not.

I will endeavor to trace a coherent path from the beginning of the developmental process (i.e., insects such as the silkworm were subject to disease caused by *B. thuringiensis* to the present state of development (i.e., *B. thuringiensis* can be grown in quantity and distributed in the field as an effective biocontrol agent). Table 1 gives the subspecific divisions of *B. thuringiensis* based on H-antigens into serotypes. The nearly 400 known isolates can be placed in one or another of these serotypes or varietal groupings. By an irony of fate, the oldest known isolate (var. *sotto*) is not accorded the place of honor as serotype 1.

TABLE 1. Subspecific divisions of *Bacillus thuringiensis* based on H-antigens into serotypes.

Serotype H[a]	Biotype	Variental epithets
1	I	*thuringiensis*
2	II	*finitimus*
3a	III 1	*alesti*
3a, 3b	III 2	*kurstaki*
4a, 4b	IV 1	*sotto*
4a, 4b	IV 1'	*dendrolimus*
4a, 4c	IV 2	*kenyae*
5a, 5b	V 1	*galleriae*
5a, 5c	V 2	*canadensis*
6	VI	*subtoxicus*
6	VI 1	*entomocidus*
7	VII	*aizawai*
8a, 8b	VIII 1	*morrisoni*
8a, 8c	VIII	*ostriniae*
9	IX	*tolworthi*
10	X	*darmstadiensis*
11a, 11b	XI 1	*toumanoffi*
11a, 11c	XI 2	*kyushuensis*
12	XII	*thompsoni*
13	XIII	*pakistani*
14	XIV	*israelensis*
15	XV	*dakota*
15	XVI	*indiana*
-	-	*wuhanensis*[b]

[a]Based on classification according to Dr. H. deBarjac, Institut Pasteur, Paris, France.

[b]No flagellar antigen.

In early Japanese texts on silkworm culture, clear evidence has been presented to demonstrate that periodic cleaning of silkworm rearing facilities was crucial to the successful production of silk. Emphasis was placed on the removal of debris accumulated in the rearing facility and the necessity of annual fumigation. Nevertheless, the first isolate of *B. thuringiensis* was described by the Japanese. In a series of papers, Ishiwata (1901, 1905) described a species of bacteria isolated from diseased silkworm and he expressed the opinion that the isolated organism could be one of the causes of flacherie (a catch-all descriptor applied to many infectious lethal diseases). He named this microorganism "*Sotto bacillus*" since infected larvae "very suddenly fall over on their sides and die" (*sotto* in Japanese means "sudden collapse"). Ishiwata found that suspensions of old agar cultures (7 days - 9 months old) were lethal when ingested by silkworm larvae. Filtrates of broth cultures were not toxic.

Aoki and Chigasaki (1915a, 1916a,b) continued the work of Ishiwata with the *sotto* bacillus and reported that only old agar cultures were pathogenic when ingested by silkworm. They noted that the lethal action was not correlated with the multiplication of ingested bacteria since young vegetative cells were destroyed in the insect gut. Silkworm larvae died when *sotto* was injected into the body cavity and young cultures were more virulent by this method than older cultures. These results led them to conclude that the larvae were killed through the action of a substance that was already present in old agar cultures. After investigating the toxicity of sporulated agar cultures, they found that the toxic principle was not present in the medium in which the bacteria were grown but was associated with the bacterial colony. The substance responsible for the toxic effect was not filterable through clay candles, and boiling for 10 min destroyed its toxic activity, as did iodine, formalin, mercuric chloride, and alcohol. They also found that the agglutination reaction for *sotto* was very specific. Through use of this reaction it was demonstrated that the *sotto* bacillus could be easily distinguished from *B. megaterium* and *B. alvei*.

With the ponderous assurance of hindsight (and the advantage of 60 years of progress in bacteriology, biochemistry, and insect physiology) it is evident that these early Japanese workers had discovered all the essential features of the *B. thuringiensis* story: sporulated cultures of the *sotto* isolate contained a toxic material capable of killing silkworm when ingested. Curiously, the idea of *sotto* being used as a microbial insecticide was not evoked by these investigators.

B. thuringiensis is not a highly infectious pathogen and under natural conditions, does not cause a great deal of mortality. The most likely explanation would seem to lie in the fact that cadavers of insects killed by it contain enormous numbers of vegetative cells but relatively few spores and crystals. In other words, the principal agents of mortality do not build up rapidly in an insect population. The fact, however, that most of the known varieties of *B. thuringiensis* have been isolated from insects argues that it is a successful parasite, if persistence (even at a low level) is a measure of success.

Provided sufficient nutrient is available, vegetative cell multiplication of *B. thuringiensis* can occur at low oxygen levels but sporulation and crystal production occur best with vigorous aeration of liquid cultures, or on solid media where the surface growth is exposed to atmospheric oxygen (Angus, 1968). Of course, such conditions do not occur in an infected insect. On first gaining access to the hemocoel, vegetative cells grow vigorously at the expense of the host tissue, adding an additional oxygen requirement to their host (Angus, 1968). As the insect becomes moribund, the oxygen level in the hemolymph and tissues decreases because diffusion through the tracheoles is normally assisted by the movement of the insect. Thus, a falling available nutrient level is accompanied by a falling oxygen level; the one process triggers spore (and crystal) formation, the other operates against it. Obviously, enough spores are produced to ensure persistence of the bacterium because it does persist, albeit at low levels, as evidenced by its periodic isolation from granaries where the flour moth is present, and in commercial cocooneries.

It may seem anomalous to use a relatively non-infectious pathogen as a microbial insecticide. However, culturing of *B. thuringiensis* under highly artificial conditions (constant temperature, pure culture, optimum nutrient levels, abundant and constant oxygen levels, etc.) makes it possible to obtain material that has a much higher count of spores and toxic crystals than would be otherwise possible. The distribution of such material in the field exposes insects to a risk of infection very much higher than would ever occur under natural conditions (Angus, 1968).

It should be emphasized that Aoki and Chigasaki (1915a,b; 1916a,b) were probably working exclusively with silkworm. Had they the opportunity to test other Lepidoptera, they might have been alerted to the fact that strains of *B. thuringiensis* affect many lepidopteran larvae. In any event, investigations of the *sotto* isolate languished. Fortunately, the *sotto* isolate was placed in a stock culture collection and faithfully transferred. Later, its potential would be realized as a disenchantment with chemical insecticides was voiced and its importance became attractive to funding agencies interested in the development of alternate means for insect control.

The next thread in our story occurred in the Thuringia region in central Germany. Indeed, the name still persists today to describe the people of the area and to describe a particular kind of sausage having its origin here. As we all recognize, a derivation of Thuringia also identifies a very important sausage-shaped *Bacillus*. It was in Thuringia that Berliner (1911, 1915) found diseased and dead larvae of the Mediteranean flour moth in a flour mill and isolated a spore-forming bacterium that he called *B. thuringiensis* in recognition of its origin. Berliner's work was highly noteworthy because not only did he isolate the causative agent of the disease he observed, but he also demonstrated that the organism could induce new infections of healthy larvae from fresh cultures when given per os. Berliner noted that after sporulation, rhomboidal-shaped structures could be identified in the debris. Athough Berliner speculated that *B. thuringiensis* could be useful in biological control no development of his suggestion was initiated. One should recall however, that Berliner's final paper on *B. thuringiensis* appeared in 1915 and by then World War I had been launched; destruction of insects had become of secondary importance.

I will comment only briefly on the work with *B. thuringiensis* that was sponsored by the International Corn Borer Investigations (ICBI) Research Program in Central Europe. The program was carried out during 1928-31 by European scientists with the help of North American funding. Much of their findings was published in a series of very brief annual reports (Metalnikov et al., 1928-1931) and not in scientific journals. Reading these reports with the advantage of hindsight reveals some intriguing facts in their discussion of new isolates of spore-formers pathogenic to corn borer larvae:

> "*Bacterium pyrenei* No. 1 is a long spore-forming Gram positive rod. The oval spores are never formed in the middle but always toward the end of the rods. It is one of the most virulent bacteria we possess. Corn borer larvae and larvae of other Lepidoptera often die in 10-15 hours when infected *per os* by a virulent strain. The dead larvae always take on a deep black colour...*Bacterium italicam* No. 2 - The spores are formed towards the ends of the rods".

It was also noted that *B. thuringiensis* is very virulent against the gypsy moth and some other lepidopteran larvae but harmless to grasshoppers, mosquitoes, and beetles. Also included in the reports was the tantalizing clue "it is interesting to note that the old bacterial cultures are the most virulent". Lastly, it was stated that *Bacterium ephestiae* No. 1 and No. 2

should be identified as *B. thuringiensis* and that the spores are formed near one end of the rod.

Although the I.C.B.I. reports were consistently positive in tone, by reporting good progress and by expressing interest in proceeding with further trials of potentially useful strains of microorganisms, the 1931 report was the last to appear and the project ceased to exist. The project was terminated presumably because of the difficulty of securing funding as the Great Depression deepened and the fact that the corn-borer became less urgent as a problem. From our point of view, the I.C.B.I. work established that *B. thuringiensis* (and related strains) were pathogenic for many lepidopterous pests and could effect modest population reductions when released in the field. In retrospect, it is realized that the I.C.B.I. group was searching for highly infectious agents; our contemporary experience indicates that the successful use of *B. thuringiensis* depends on its application as an inundative agent. Unfortunately, most of the I.C.B.I. isolates were not maintained; there is little doubt, based on inferential evidence, that many of thse isolates would be identified today as serotypes of *B. thuringiensis*.

Approximately 12 years after Berliner (1915) described the type species of *B. thuringiensis*, Mattes (1927) re-isolated the organism and described the symptomalogy evoked in diseased flour moth larvae. The most interesting of the findings was a description of Berliner's rhomboidal bodies as "Restkorper" (loosely, remaining-behind-bodies); no function was assigned to the structures. Fortunately, the Mattes' isolate was preserved. It passed through many laboratories finally lodging with Edward Steinhaus in 1942. After remaining for 7 years as a "Cinderella" organism, Steinhaus (1951) and his colleagues discovered that sporulated cultures of *B. thuringiensis* were active against the alfalfa caterpillar. The notion of using *B. thuringiensis* as microbial insecticide began to be more actively considered.

Meanwhile, Toumanoff and Vago (1951, 1952, 1953a) published a number of papers devoted to the various aspects of studies on *B. cereus* var. *alesti*, the causal agent of an endemic flacherie in French silkworm populations. The symptoms described in silkworms infected with this organism commence with sluggishness, than partial paralysis followed by total paralysis, and finally death. Toumanouff and Vago distinguished two modes of action in their reports. The first involved a rapid lethal action (toxemia) when the larvae ingested large doses of old cultures. The second was described as slow morbid action invariably accompanied by septicemia when the larvae ingested small doses.

In a histological study of larvae infected with *B. cereus* var. *alesti*, Toumanoff and Vago found the same effects described for the *sotto* strain by Aoki and Chigasaki (1915a,b; 1961a,b). In fact, when toxemia was the primary cause of death, ingested spores had not germinated in the gut of the killed larvae.

Noting that the identity of *Bacillus sotto* had never been precisely established, Toumanoff (1952) initiated a comparative study of the Japanese strain of *B. sotto* and two strains of *B. thuringiensis* known to be pathogenic for some lepidopterous larvae. He found that *B. sotto* was closely identical to *B. thuringiensis*. Later, Toumanoff and Vago (1953b) concluded that the toxicity of *B. cereus* var. *alesti* was the result of a "substance elaborated in the course of bacterial metabolism which fixes itself on the bacterial cell and is found probably with the spores".

In Canada, at the Sault Ste. Marie Laboratories, Art Heimpel and myself had unknowingly repeated most of the Japanese work with the *sotto* isolate and the silkworm, *Bombyx mori*. We were able to demonstrate that the symptoms evoked in silkworm larvae after ingesting sporulated cultures of *B. thuringiensis* had nothing to do with spore germination and vegetative cell multiplication in the insect but rather an effect more similar to poisoning or toxinosis (Angus and Heimpel, 1960; Heimpel and Angus, 1960, 1963). When washed sporulated cultures were homogenized, ingestion of the non-viable paste evoked the characteristic symptoms - immediate cessation of feeding followed by a progressive general paralysis.

Thus by 1954 nearly all the pieces of the *B. thuringiensis* puzzle were in place: (1) young cultures of a closely related group of spore-forming bacteria when ingested caused little mortality to susceptible caterpillar larvae, but old or sporulated cultures were often quite pathogenic *per os*; (2) supernatant fluids of liquid cultures were not pathogenic; (3) the displacement of the spore organism to one end of the sporangium occurs when the crystal is formed; (4) the cellular debris of lysed cultures contained rhomboidal bodies; (5) only lepidopteran species seemed to be susceptible to the pathogen; (6) the toxic principle was heat labile and could be denatured by various chemicals; and (7) death of the susceptible insect larvae usually occurred before there was detectable outgrowth of the ingested spores in the gut.

The final element in the explanation was provided when Dr. Edward Steinhaus sent to Dr. Carl Robinow, a microbiologist well known for this pioneering work on bacterial cytology, a culture of *B. thuringiensis* and requested his opinion on the "cocked-to-one-side" or "pushed to one end" position of the spore noted in the isolate. Dr. Robinow subsequently consulted with Dr. Hannay of the Canada Agriculture Laboratory in London, Ontario, about the problem. Following some very elegant microscopy and culture studies, Hannay (1953) showed unequivocally that the Mattes Restkörper was a crystalline parasporal-inclusion body. In addition, he suggested that the crystals could be the source of the insecticidal activity observed by our research group. We later demonstrated at the Sault Ste. Marie Laboratory, Canada, that the parasporal crystals were proteinaceous and that pure preparations or digests of the crystals evoked the characteristic symptoms in the silkworm *Bombyx mori*, and to a lesser degree similar symptoms in other lepidopteran larvae (Heimpel and Angus, 1963).

Through Edward Steinhaus' enthusiastic missionary work directed at the promotion of *B. thuringiensis* as a biocontrol agent, industry became interested in its potential and in November of 1956 the first commercial product of *B. thuringiensis* and *B. sotto* were available for testing. The next year in a mimeographed literature survey, Steinhaus reported that 54 species of Lepidoptera had been found susceptible to *B. thuringiensis*, 10 species to the *sotto* isolate, and 10 species to the *alesti* isolate.

In retrospect, once parasporal crystal function had been explored and reasonably understood, many perplexing anomalies were resolved which allowed industry to optimize the production process, thus *B. thuringiensis* moved from the bench to the field on a rational basis.

III. CONCLUSIONS

Professor Ed Steinhaus (1967) in a retrospective review stated:

> "I believe there is a place in all this for a note of genuine humility. I have frequently said that our Laboratory has received far more than its share of credit in the development of crystalliferous bacteria as microbial insecticides. This is said as no idle platitude. Rather, it springs from the conviction that all of us owe so much to those who have preceded us, to those who have taught and inspired us, to our contemporaries in other laboratories and other countries, and to our close associates, that to assume our own contributions paramount is not only ungracious, but lacking in objectivity and scientific realism.
>
> I must make the point, no doubt an obvious one, that what is now a commercial reality in the form of a microbial insecticide, did not become this

> overnight. Years of basic research preceded and accompanied the practical utilization of *B. thuringiensis*. We should not be allowed to forget that these rewards arise out of the curiosity of the mind of man and his desire to seek knowledge for knowledge's sake."

It was characteristic of Ed Steinhaus that he would depreciate his own great contributions, applaud those of his fellows, and remind us all of our common inter-dependence.

I hope the reader will forgive me if I end on a somewhat personal note. I have deeply appreciated this opportunity to honor both Ed Steinhaus and Art Heimpel. They were close friends of mine and we had fun working together, both in science and in the affairs of the Society for Invertebrate Pathology. In a few weeks, I will be taking a shorter haul on the rope by retiring, but I will always treasure these times we have had together. Thank you for everything -- the chance to work and play together, and the gift of your friendship.

IV. REFERENCES

Angus, T. A. (1968). The use of *Bacillus thuringiensis* as a microbial control agent. *World Rev. Pest Control*, 7, 11-26.

Angus, T. A. (1970). Implications of some recent studies of *Bacillus thuringiensis* - A personal purview. *In* "Preceedings IV International Colloquium on Insect Pathology", College Park, MD. pp. 183-189.

Angus, T. A. and Heimpel, A. M. (1960). The bacteriological control of insects. *Proc. Entomol. Soc. Ontario*, 90, 13-21.

Aoki, K. and Chigasaki, Y. (1915a). Ueber das Toxin von sog. Sotto-Bacillen. *Mitt. Mediz.* Fakult. Kais. Univ. Tokyo, 14, 59-80.

Aoki, K. and Chigasaki, Y. (1915b). Ueber die Pathogenitat der sog. Sotto-Bacillen (Ishiwata) bei Seidenraupen. *Mitt. Jediz.* Fekult. Kais. Univ. Tokyo, 13, 419-439.

Aoki, K. and Chigasaki, Y. (1916a). Ueber atoxogene Sotto-Bacillen. *Bull. Imp. Ser. Exp. Stat.*, Tokyo, 1, 141-149.

Aoki, K. and Chigasaki, Y. (1916b). Ueber die Anwendbarkeit der Agglutinationsreaktion bei der bacteriologischen Untersuchung von *Bacillus sotto* (Ishiwata), *B. alvei* (Chesire et Cheyne) und *B. megaterium*. *Bull. Imp. Ser. Exp. Stat.* Tokyo, 1, 83-95.

Berliner, E. (1911). Uber der Schlaffsucht der Mehmotten raupe. *Z. ges. Getreidw.*, 3, 63-70.

Berliner, E. (1915). Ueber die Schlaffsucht der Mehlmottenraupe (*Ephestia kuehniella* Zell.) und ihren Erreger *Bacillus thuringiensis* n. sp. *Z. Angew. Entomol.*, 2, 29-56.

Chesire, F. R. (1885). The pathogenic history and history under cultivation of a new bacillus (*B. alvei*), the cause of a disease of the hive bee hitherto known as foul brood. *J. Royal Microsc. Soc.*, 5, 581-601.

Hannay, C. L. (1953). Crystalline inclusions in aerobic spore forming bacteria. *Nature*, 172, 1004.

Heimpel, A. M. and Angus, T. A. (1960). Bacterial insecticides. *Bacteriol. Rev.*, 24, 266-288.

Heimpel, A. M. and Angus, T. A. (1963). Diseases caused by certain spore-forming bacteria. *In* "Insect Pathology - An Advanced Treatise" (E. A. Steinhaus, ed.), Vol. 2, 21-73. Academic Press, New York.

Ishiwata, S. (1901). On a kind of severe flacheries (sotto disease). *Dainihon Sanshi Kaiho* (Report of the Sericultural Association of Japan), No. 114, 1-5. (In Japanese).

Ishiwata, S. (1905). About "Sottokin" a bacillus of a disease of the silkworm. *Dainihon Sanshi Kaiho* (Report of the Sericultural Assoc. of Japan) No. 160, 1-8; No. 161, 1-5.

Knipling, E. F. (1970). The role of microbial agents in the development of alternative means of insect control. *In* "Proceedings IV International Colloquium on Insect Pathology," College Park, Maryland. pp. 2-8.

Mattes, O. (1927). Parasitare Kranheiten der Mehlmottenlarven und Versuche uber ihre Verwendbarkeit als biologisches Bekamfungsmittel. *Sitzber. Ges. Befordeer, Ges. Naturw.*, 62, 381-417.

Metalnikov, S. et al. (1928-1931). International Corn Borer Investigations - Science Reports. 1 (1928), 2 (1929), 3 (1930), 4 (1931) (Extensively quoted in "Insect Microbiology" E. A. Steinhaus, Comstock, Ithaca, 1946).

Steinhaus, E. A. (1951). Possible use of *B. thuringiensis* Berliner as an aid in the biological control of the alfalfa caterpiller. *Hilgardia*, 20, 359-381.

Steinhaus, E. A. (1956). Microbial Control - The emergence of and idea. *Hilgardia*, 26, 107-157.

Steinhaus, E. A. (1967). Entomology looks at its mission. *Bull. Entomol. Soc. Amer.*, 13, 104-108.

Toumanoff, C. (1952). A propos d'un bacille pathogene pour les vers a soie au Japan (*Bacillus sotto*) et ses affinites avec d'autres bacilles entomophytes. *Ann. Inst. Pasteur*, 82, 512-516.

Toumanoff, C. and Vago, C. (1951). L'agent pathogene de la flacherie des vers a soie endemique dans la region des Cevennes: *Bacillus cereus* var. *alesti* var. nov. *C. R. Acad. Sci. Paris*, 233, 1504-1506.

Toumanoff, C. and Vago, C. (1952). La nature de l'affection des vers a soie due à *Bacillus cereus* var. *alesti* et les modalites d'action de ce bacille. *Ann. Inst. Pasteur*, 83, 421-422.

Toumanoff, C. and Vago, C. (1953a). Etude histopathologique des vers a soie atteints de *Bacillus* var. *alesti*. *Ann. Inst. Pasteur Paris*, 84, 376-388.

Toumanoff, C. and Vago, C. (1953b). Recherches sur l'effect toxique de *Bacillus cereus* var. *alesti* vis-a-vis des vers a soie. *Ann. Inst. Pasteur Paris*, 84, 623-627.

CONTROL OF THE HORN FLY, *Haematobia irritans*, WITH *Bacillus thuringiensis*

Richard E. Gingrich

U. S. Livestock Insects Laboratory
Agricultural Research Service
U.S. Department of Agriculture
Kerrville, Texas

I. INTRODUCTION

The horn fly, one of the most common pests of cattle, causes an estimated annual loss of over 350 million dollars to producers in the United States. The adult flies are hematophagous and spend most of their lives on the host where their frequent feeding causes almost constant irritation. As few as 12-22 flies/dairy animal (Granett and Hansens, 1956) or 46-59 flies/grazing beef animal (Granett and Hansens, 1957) can cause an economic loss, but hundreds per individual animal are common, and occasionally several thousand flies/animal can occur. Present control technologies are based on chemical insecticides, which must be applied repeatedly to reduce fly numbers to tolerable levels. There are no pathogens registered for use against horn flies. Three commercial preparations of *Bacillus thuringiensis*, when added to feces or fed to animals, have been found to be active against larvae of the horn fly (Gingrich, 1965). However, consecutive daily treatments were necessary to maintain complete inhibition of larval development.

Gingrich and Eschle (1966) reported that the ingredient in a commercial preparation of *B. thuringiensis* active against horn fly larvae was insoluble. Soluble products of the bacteria were also active when added to feces, but they did not pass through animals that were treated per os. Later, Gingrich and Eschle (1971) explained this observation by demonstrating that only the water-insoluble *B. thuringiensis* β-exotoxin completed passage through the digestive tract of ruminants. Spores of *B. thuringiensis* survived passage through the digestive tract, but did not proliferate in the feces to produce insecticidal activity.

II. FIELD TRIALS WITH TREATED FEED

In 1967 Gingrich and Eschle (unpubl.) conducted field experiments with an experimental product that contained enhanced amounts of water insoluble *B. thuringiensis* β-exotoxin. This product (LD_{50} 1.6 mg/100 g feces) was three times more potent than any previously tested materials. Preliminary tests demonstrated that 6 g of this new microbial product given daily to a 500-lb bovine rendered the feces completely inhibitory to development of horn fly larvae. Later, the product was incorporated into a 20% protein range cube at the rate of 6 g/lb and fed daily. Laboratory feed tests confirmed that 1-lb of cubes prevented development of horn fly larvae in the feces.

In field trials 20 pregnant cows were enclosed in a 60-acre pasture in south central Texas and fed 1-lb of cubes/head per day for 20 days. Because of a lack of natural forage, the animals were also given 2-lb of untreated range cube and 12-lb of hay/head per day. In another pasture (3 miles distance) a control herd of 18 cows were given 3-lb of untreated cubes/head per day but no hay, since range conditions were better. Fourteen days before the start of the feed tests, the fly populations on the two herds of animals were eliminated and thereby equalized through a single application of DDVP on all of the animals in both herds. On the 1st day of treatment 2500-4000 horn flies captured from cattle in nearby pastures were released on the test herd when it appeared that the fly populations were not increasing naturally as desired. Releases were repeated three times. No horn flies were released on the control herd because horn fly populations recovered naturally.

Twice each week freshly-dropped feces were collected from five animals in the test herd and from one animal in the control herd, and all samples were returned to the laboratory and bioassayed. In addition, the average numbers of adult horn flies/head were estimated twice a week for each herd. Treatments were stopped after 20 days and the results were evaluated.

About a month later, treatments were resumed at twice the original rate (i.e., at 2 lb of treated cubes/head per day). The populations of adults were not removed before the start of this test. Also, no horn flies were released. On five occasions during the hiatus between the two treatment intervals, large numbers of flies were removed from the treated herd for other purposes, although sufficient numbers were always left so the horn fly population could begin an increase. Measurements were made as described earlier. In addition three freshly-dropped fecal pats were marked where they fell for identification each week during the treatment period. After 24 hr these pats were covered with a cage that allowed collection of all adult flies.

Figure 1 indicates that during the first stage of the treatment program the feces collected from treated naimals and bioassayed in the laboratory against horn fly larvae did not completely inhibit larval development. Nevertheless, despite the frequent releases of adult flies the number of flies on the test animals did not increase when compared to the control animals. Apparently, larval development was severely suppressed. Within 3 days after the final treatment, larvicidal activity in feces dropped sharply; 12 days later, the time required for maturation of immature horn flies that oviposited after the treatments had ceased, a dramatic increase in numbers of adult horn flies on the animals occurred. Thus, *B. thuringiensis* was indicated as a suppressant for adult horn fly populations.

During the second stage of the field trials, the laboratory bioassay again showed inhibition of larvae; however, despite the increase in treatment levels, complete inhibition was still not achieved. However, when feces was left in the field, there was no larval development. Undoubtedly, natural climatic and biotic factors contributed to the effectiveness of the field treatment. The number of adults on treated animals varied but in general, they were lower than the number of horn flies on the untreated animals. The populations of flies on the treated animals were likely supported by immigrants from the very large populations occurring on the control animals 3 miles away. With the onset of fall weather, the population of flies on both herds disappeared by the end of the 1st week in November.

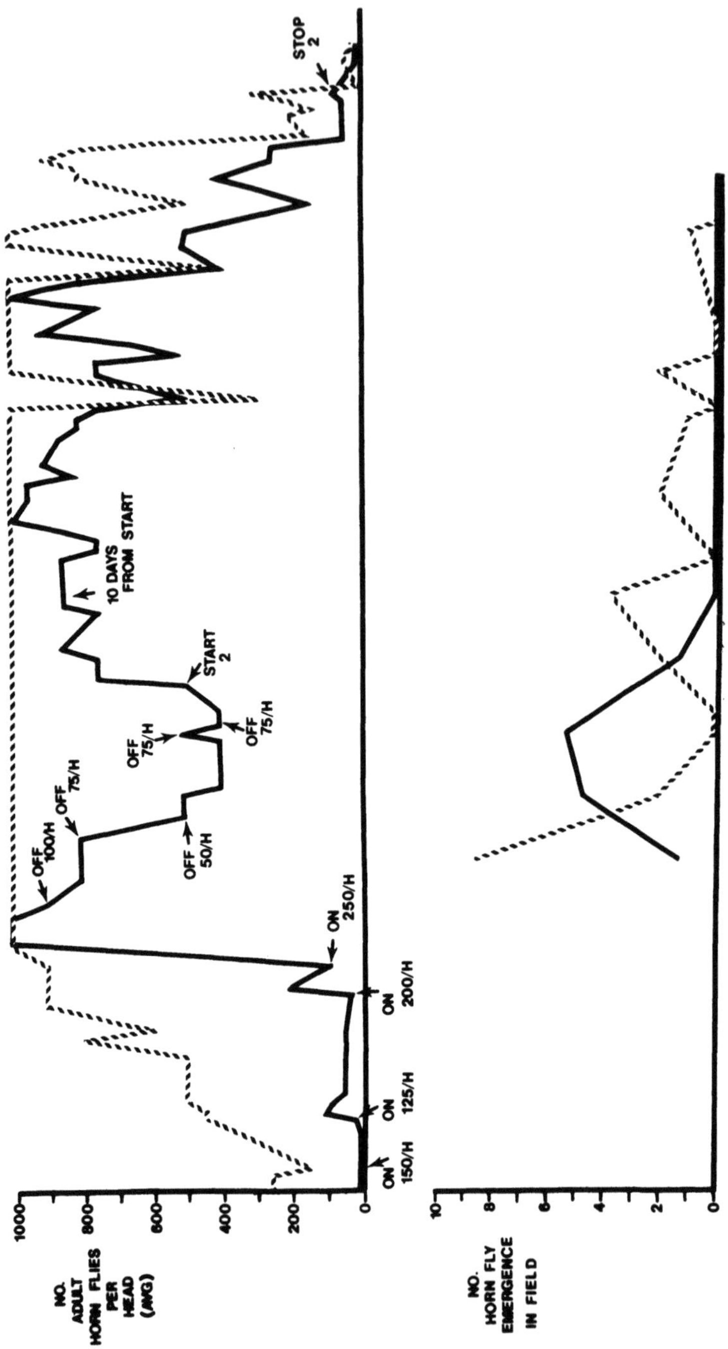
NO.
ADULT
HORN FLIES
PER
HEAD
(AVG)
1000
800
600
400
200
0
ON
150/H
ON
125/H
ON
200/H
ON
250/H
OFF
100/H
OFF
75/H
OFF
50/H
OFF
75/H
OFF
75/H
START
2
10 DAYS
FROM START
STOP
2
NO.
HORN FLY
EMERGENCE
IN FIELD
10
8
6
4
2
0

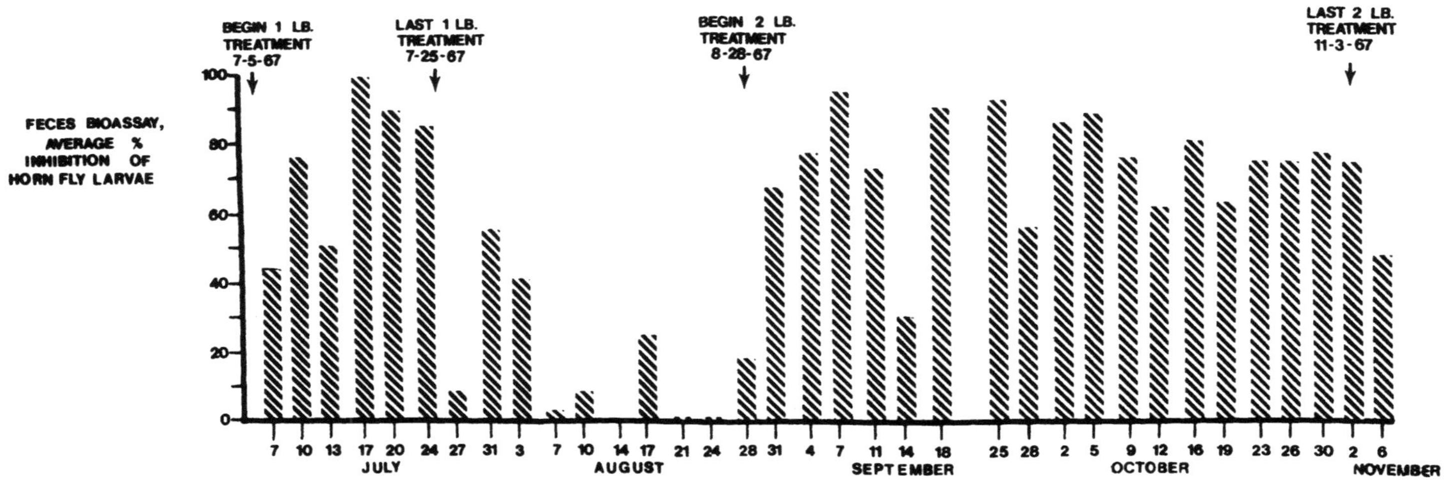

Figure 1. Feces bioassay, horn fly emergence in field, and average number of horn flies on individual animals in a field study to determine the effect of feeding range cubes treated with β-exotoxin of *Bacillus thuringiensis* to a beef herd. Solid line (-) indicates data for the treated herd; broken line (---) indicates data for the untreated herd.

There were no measurable ill-effects on the 20 pregnant cows that could be attributable to the *B. thuringiensis* β-exotoxin either during or after the treatment period. The 85% calving record of the following spring was similar to previous fecundity data in both test and control herds. One animal delivered a stillborn, deformed calf, which also occurs occasionally among untreated herds.

After the field test and approximately 7 months after their formulation, the range cubes were bioassayed in a laboratory feed test with steers. The results indicated that the potency had not decreased. In another follow-up experiment, 1/2 lb of treated cubes were fed for 10 days to one pregnant cow and two "dry" cows under controlled laboratory conditions. Bioassay of the feces revealed that treatments were equally effective for all animals and that pregnancy did not reduce the effectiveness of the *B. thuringiensis* β-exotoxin.

In summary, the *B. thuringiensis* β-exotoxin formulated into range cubes and fed to cows in the field gives effective inhibition of horn fly larvae development in the feces. When populations of adult flies are low, inhibition by the *B. thuringiensis* β-exotoxin, in addition to natural inhibitory factors, is sufficient to retard or prevent the normal increase of horn fly adults during their breeding season.

III. EPIZOOTIC IN A LABORATORY COLONY

Until recently, only heat-stable toxins of *B. thuringiensis* was reported as effective against species of Diptera. However, because of evidence that the heat-stabile β-exotoxin was pathogenic for mammals (Sebesta and Horska, 1968; de Barjac and Riou, 1969; Sebesta et al., 1969), a few attempts have been made to commercialize these toxins. No heat-labile components of the bacterium were known that were pathogenic to these insects until Hall et al. (1977) demonstrated that the heat-labile fractions of several strains of *B. thuringiensis* were active against mosquitoes.

In July, 1975, I began investigating the cause of a steadily declining production in a mass-rearing colony of the horn fly at the U.S. Department of Agriculture's Livestock Insects Laboratory, Kerrville, Texas. Adults produced from the mass-rearing program were used for release in an eradication project. The rearing facility was designed to closely accomodate the natural breeding instincts of the insect (Berry et al., 1974). For example, adults were released and fed on several bulls that were held in stanchions in a steel building, and gravid females oviposited on feces dropped behind the animals. The infested feces were collected once a day, mixed with water and dry ingredients, including begasse fibers, and then held in the same building for 3-5 days until completion of pupation. The rearing medium was transferred to chambers in a separate building 50 miles away where adult eclosion took place. Adults were either irradiated for release in the field or were untreated and released on the host animals in the rearing building. Production approximated 300,000 flies/day before the decline began.

Initial quality control checks by the rearing personnel indicated that the greatest loss was among larvae in the rearing containers. Upon examination, I found several dead third stage larvae on the surface of the rearing medium. They were disinfected externally and cultured on nutrient agar; some colonies were identified as *B. thuringiensis*. I will now trace chronologically the events that finally culminated in an epidemic of serious consequences to the insect colony.

On the first day of my investigation *B. thuringiensis* was isolated from additional dead larvae, but survival of adults on the bulls, egg hatch, and emergence of pupae were normal. Since the problem seemed to be limited to larvae and because of the reputed action of *B. thuringiensis* against horn fly larvae, I assumed plant products contaminated with *B. thuringiensis* β-exotoxin-producing strain of *B. thuringiensis* were probably being fed to the cattle that produced the feces used in the rearing medium (or some similarly contaminated materials were being used for the additives to the rearing medium). Therefore, feces from a bull that was notable for low production of insects were cultured on day 4 and found to be positive for the presence of *B. thuringiensis*. During the next few days, repeated checks for the presence of *B. thuringiensis* were made on all the ingredients used in the rearing medium, on all materials fed to the animals, on unemerged pupae from the emergence chambers, and on samples of adult horn flies removed from the animals. Also, daily checks of feces from each animal were begun. More isolations of the bacteria, still restricted to larvae and feces, supported my original diagnosis of the problem. There was no evidence to suggest that the bacterium was infesting any stage of the insect or that it was being transmitted wihin the insect's life cycle.

New supplies of animal feed and ingredients for the larval medium were obtained, checked to insure they were not contaminated, and then used in place of the original materials. Nevertheless, the problem with larval mortality persisted. Then, on day 8 *B. thuringiensis* was isolated from adult flies taken from the same animal whose feces had originally harbored the organism. Because the adults could have contracted the organism from the feces during oviposition, there was still no real evidence of infection within the colony. However, on day 18 and again on several subsequent occasions during the next 12 days, *B. thuringiensis* was isolated from the internal contents of puparia. Thus, it now appeared that the bacterium was infecting the insects and that vertical transmission between stages was occurring. Additional evidence to support this hypothesis appeared on day 34 with the isolation of *B. thuringiensis* from newly emerged adults. As a result, the insect colony was destroyed, the bulls were turned out, and the building and contents were cleaned, disinfected with chemicals, and fumigated with methyl bromide. Feces from the bulls that were now held in outdoor pens were checked for 9 more days with *B. thuringiensis* being isolated intermittently until the last day of sampling.

Four days after the rearing facilty was fumigated seven different bulls were installed and infested with adult horn flies that were captured on cattle about 50 miles away. After a few more days, when pupae had completed development, the emergence facility was also disinfected and fumigated, and the events calendar was set back to day 0. The feed and medium ingredients for larval rearing were checked in addition to the feces from the new animals. These samples were found to be free of *B. thuringiensis*. However, on day 13 the organism was isolated for the first time from feces of two of the new animals. On the next day, a heavy infestation of mites on the larval-rearing medium was noticed. Additional isolation of *B. thuringiensis* were made from feces in the following days, and by day 23 *B. thuringiensis* was isolated from horn fly pupae. Meanwhile, the problem with the mites increased as species of Pyemotidae and Anoetidae were found in the larval medium and on the adult flies. Miticides were applied unsuccessfully in an attempt to prevent their spread to newly-prepared larval rearing containers.

On day 35 adult horn flies infested with mites were captured and the mites removed from their hosts. The internal contents of the mites were cultured and *B. thuringiensis* was isolated. Two days later, mites collected from newly emerged flies in the emergence chamber also were found to be contaminated with *B. thuringiensis*. Further isolations of the bacterium from larvae, pupae, and adults of the insects, from feces of their hosts, and from two species of mites that infested the adult flies and the larval medium were achieved. Production of horn flies continued to decline until there were no longer adequate numbers for irradiation and release. Indeed, fly production had dropped 90% in less than 2 months. Finally, in December, 1975, the epizootic of *B. thuringiensis* could no longer be controlled and the colony was terminated. Furthermore, the success of the pilot eradication project for the horn fly had been jeopardized by infection of the supporting insect colony by *B. thuringiensis*. Not only the quantity, but also the quality, of the insects produced was impaired.

Etiology and Diagnosis. During the epizootic, there were more than 50 separate isolations of *B. thuringiensis*. To determine their role in the destruction of the colony it was necessary to determine how many strains of the organism were present and whether they were pathogenic for horn fly larvae. The subsequent biochemical and morphological studies indicated that at least 7 serotypes (1, 2, 3ab, 5ab, 6, 8, and 13) of *B. thuringiensis* were present.

Selected isolates were grown in submerged culture and bioassays were performed on normal and autoclaved samples of the supernatant fluids and sediment fractions. Again, the results showed the strain heterogeneity of *B. thuringiensis* isolated from the insect colony. As expected, some isolates, particularly those found in larvae during the early stages of the epizootic, belonged to serotype 1 and produced the *B. thuringiensis* β-exotoxin. Curiously, some strains produced heat-labile insoluble agents that killed horn fly larvae. These latter isolates appeared coincidentally with the discovery of infected pupae and adults and were identified as serotypes 3ab and 5ab. Thus, there was strong evidence that spores and the δ-endotoxin of some strains of *B. thuringiensis* were pathogenic to the horn fly. This finding supports the earlier speculation that a possible bacterial infection was being vertically transmitted in the mass-rearing colony.

The more recent studies in our laboratory indicate that at least one isolate in serotype 3ab produces crystals that are toxic to horn fly larvae and that another isolate in serotype 5ab produces spores that have pathogenicity to the horn fly larvae. Studies with these two isolates as well as other select isolates known to produce the *B. thuringiensis* β-exotoxin are continuing in an attempt to develop biological control agents that can be used against horn flies and possible other Diptera attacking livestock.

Epidemiology. The original source of the *B. thuringiensis* contamination in the horn fly colony is unknown. Several years ago, commercial preparations containing serotype 1 strains of *B. thuringiensis* were tested in the area by feeding them to livestock; however, none of the preparations were used in the building where the epizootic occurred. Perhaps, as originally suspected, food provided to the host insects was contamined, although a positive identification was never made. Of course, personnel from the insect pathology laboratory could have inadvertently carried the bacterium into the rearing facility during brief visits. Laboratory personnel were working at the time with many isolates of *B. thuringiensis*, including some that Hall et al. (1977) found active against mosquitoes. However, the strains that Hall and his coworkers found most active were not pathogenic to horn flies. Also, some of the isolates from the diseased insect colony were unique in identity and character from the isolates maintained in the insect pathology laboratory. Thus, although some of the strains isolated from the colony were possibly introduced by personnel from this laboratory, those strains that were pathogenic and apparently responsible for the loss of the colony were not.

The re-infestation of the facility after it was cleaned, disinfected, stocked with new insects and host animals, and henceforth not visited by personnel from the insect pathology laboratory suggests that sources and vectors of disease other than the obvious ones mentioned may have been involved. For example, the mites which were undoubtedly present long before the dramatic rise in the incidence of the disease, became very abundant during the second rearing attempt, and infection among the horn fly pupae and adults coincided with the isolation of identical serotypes of *B. thuringiensis* from the mites. Thus, it is possible that mites (or some other organisms) served as a vector. Still, the source of these bacteria is unknown. Clearly, many questions on the epizootiology of *B. thuringiensis* in the pathology of Diptera remain unanswered. The answers to these questions are important for success in mass-rearing insects and for development of this microbe as a biological control agent.

IV. CONCLUSIONS

Commercially prepared β-exotoxin produced by certain strains of *B. thuringiensis* incorporated into 20% range cubes at the rate of 6 g/lb effectively controlled populations of horn flies, *Haematobia irritans*, on cattle when fed in the field at the rate of 1-2 lb/head per day. The treatment had no apparent ill-effect on the 20 pregnant cows during or after the 97-day treatment period. When an epizootic of *B. thuringiensis* resulted in the destruction of a large-scale rearing colony of horn flies, several strains of the bacterium that included serotypes 1, 2, 3ab, 5ab, 6, 8, and 13, were isolated from diseased horn fly larvae, pupae, and adults, from the feces of cattle used as the larval-rearing medium, and from species of pyemotid and anoetid mites associated with the insect colony. Bioassays indicated that active heat-stable (β-exotoxin) and heat-labile fractions were produced by some of the *B. thuringiensis* isolates that have potential for biological control of the horn fly, *H. irritans*.

V. REFERENCES

Barry, I. L., Harris, R. L., Eschle, J. L., and Miller, J. A. (1974). Large-scale rearing of horn flies on cattle. *USDA Publication ARS-S-35* (January, 1974).

deBarjac, H. and Riou, J. Y. (1969). Action de la toxine thermostable de *B. thuringiensis* administree a des souris. *Rev. Pathol. Comp. Med. Exp.*, 6, 367-374.

Gingrich, R. E. (1965). *Bacillus thuringiensis* as a feed additive to control dipterous pests of cattle. *J. Econ. Entomol.*, 58, 363-364.

Gingrich, R. E. and Eschle, J. L. (1966). Preliminary report on the larval development of the horn fly, *Haematobia irritans*, in feces from cattle given fractions of a commerical preparation of *Bacillus thruingiensis*. *J. Invertebr. Pathol.*, 8, 285-287.

Gingrich, R. E. and Eschle, L. L. (1971). Susceptibility of immature horn flies to toxins of *Bacillus thuringiensis*. *J. Econ. Entomol.*, 64, 1183-1188.

Granett, J. and Hansens, E. J. (1956). The effect of biting fly control on milk production. *J. Econ. Entomol.*, 49, 465-467.

Granett, J. and Hansens, E. J. (1957). Further observations on the effect of biting fly control on milk production in cattle. *J. Econ. Entomol.*, 50, 332-336.

Hall, I. M., Arakawa, K. Y., Dulmage, H. T., and Correa, J. A. (1977). The pathogenicity of strains of *Bacillus thuringiensis* to larvae of *Aedes* and *Culex* mosquitoes. *Mosq. News*, 37, 246-251.

Sebesta, K. and Horska, K. (1968). Inhibition of DNA-dependent RNA polymerase by the exotoxin of *Bacillus thuringiensis* var. *gelechiae*. *Biochem. Biphys. Acta.*, 169, 281-282.

Sebesta, K., Horska, K., and Vankova, J. (1969). Inhibition of de novo RNA synthesis by the insecticidal exotoxin of *Bacillus thuringiensis* var. *gelechiae*. *Collect. Czech. Chem. Commun.*, 34, 1786-1791.

Bacillus thuringiensis: INTEGRATED CONTROL - PAST, PRESENT, AND FUTURE

Y. Tanada

Department of Entomological Sciences
University of California
Berkeley, California

I. INTRODUCTION

In this paper, I shall consider the integration of *Bacillus thuringiensis* by man in microbial, biological, and chemical control programs, i.e., in pest management. The discussion will consider the question: Is *B. thuringiensis* a prime candidate for integrated control? The answer is yes. *B. thuringiensis* has high pathogenicity for target pests; it is safe for most nontarget organisms, including humans; it is compatible with other insects pathogens; and it is compatible with chemical insecticides, miticides, fungicides, and adjuvants. I shall review past and present accomplishments and discuss future objectives. The past will cover the period up to 1960 and the present from 1960 to 1980.

II. PAST ACCOMPLISHMENTS

When Ishiwata (1901) and Berliner (1915) isolated *B. thuringiensis*, they were impressed by its high virulence and pathogenicity. Ishiwata called the bacillus, isolated from the silkworm, *sotto*-bacillus, the sudden-collapse bacillus. Berliner (1915) noted that *B. thuringiensis* isolated from *Anagasta kuehniella* was safe for vertebrates and recommended its use against the mealworm. He reported that the bacillus did not infect certain insects, such as beetles. Later, Mattes (1927) reisolated *B. thuringiensis* from the mealworm. None of these workers made any attempt to use the bacillus for insect control.

Husz (1929) conducted the first field test with *B. thuringiensis* under a very extensive program developed to control the European corn borer, *Ostrinia nubilalis*. This program was organized in 1927 by the International Life Stock Exposition and financed by industrial corporations in Chicago (Ellinger, 1928) because the corn borer was considered a serious menace to beef production.

The corn borer control program included chemical, biological, and microbial control, the ecology of the corn borer, and host plant resistance. Ten countries were involved, mostly in Europe, the United States, and Canada. This program is among the earliest attempts at integrated control and insect pest management. The pathogens that offered promise were mainly bacteria and fungi. The most promising bacterium was *B. thuringiensis*. Unfortunately, the project was dropped after four years before it could develop practical measures.

Husz (1928) obtained his culture from Otto Mattes. He found *B. thuringiensis* highly infectious for the corn borer and obtained favorable results in applying the bacillus in the field. Chorine (1930), Metalnikov and Chorine (1929a), and Metalnikov et al. (1930a,b, 1931) confirmed the effectiveness of the bacillus. A mixture of *B. thuringiensis* and other bacteria was also tested on the corn borer, but did not enhance the control obtained with *B. thuringiensis* alone (Metalnikov and Chorine, 1929a). The studies of Metalnikov and Chorine (1929b) extended the range of insect pests susceptible to *B. thuringiensis*, and also reported that certain insects were resistant, such as grasshoppers, mosquitoes, etc.

In spite of the promising results from 1929 to 1932, the use of *B. thuringiensis* was abandoned. The bacillus languished in a culture collection until Professor Edward Steinhaus resurrected it and aroused international interest, particularly the microbiological industries, in the potential of *B. thuringiensis* as a microbial control agent (Steinhaus, 1956a,b). He obtained a culture of the bacillus in 1942 from Nathan Smith who received it from J. R. Porter, who in turn had received from O. Mattes in 1936 (Steinhaus, 1975). When Steinhaus (1975) first tested the bacillus on the alfalfa caterpillar, he was so struck by its high virulence and pathogenicity that he suspected a potent chemical insecticide had contaminated his test. Steinhaus and his colleagues also developed methods for mass propagation and application of *B. thuringiensis* (Steinhaus, 1951b; Thompson, 1958). He recommended the use of the bacillus in integrated control with chemical insecticides and with insect parasites, predators, and other pathogens.

Steinhaus was the first human volunteer to expose himself directly to *B. thuringiensis* by drinking a suspension of the bacillus. I recollect Professor Steinhaus carefully instructing his technicians to take thorough notes of signs and symptoms that might develop on himself. He experienced no ill effects from the exposure.

For control of the alfalfa caterpillar, Steinhaus (1951a,b) suggested the use of a mixture of *B. thuringiensis* and a nuclear polyhedrosis virus. The bacillus would provide a rapid kill and the virus would bring about a more prolonged control. Thompson (1958) tested the mixture and found that while insect mortality was more quickly obtained with the mixture, the control of the alfalfa caterpillar was not as complete or long-lasting as that obtained with the virus alone. He recommended the use of the bacillus only in cases where the caterpillars were too near maturity for satisfactory control with the virus.

The microbiological industry, following Steinhaus' suggestion, began studies not only on the host range and application of *B. thuringiensis*, but also on the safety of the bacillus (Fisher and Rosner, 1959). Their efforts aided in the granting of a registration by the U. S. Government permitting the use of *B. thuringiensis* for insect control in the field.

During the mid 1950s, Carl Reiner and I ran a series of tests with *B. thuringiensis* for control of the plume moth (*Platyptilia carduidactyla*) on artichoke and of the corn earworm (*Heliothis zea*) on sweet corn (Tanada and Reiner, 1960, 1962). In these tests, *B. thuringiensis* was compared with parathion used for the control of the artichoke plume moth and DDT used for the control of the corn earworm. In these studies, the artichoke plants treated with parathion developed exceedingly high populations of parthion-resistant peach aphid, *Myzus persicae*; the corn plants treated with DDT had severe outbreaks of DDT-resistant two-spotted spider mite, *Tetranychus urticae* (unpub.). Adjacent plots treated with *B. thuringiensis* had significantly lower populations of these pests, indicating that the bacillus did not upset the natural enemies.

During the latter half of the 1950s, tests were conducted on the effect of *B. thuringiensis* on insect parasites and predators. For example, Biliotti (1956a,b) tested the bacillus on parasites of the European cabbageworm, *Pieris brassicae*. If the bacillus did not kill the host larvae prior to a certain stage in the development of the parasites, i.e., the third instar for the braconid and the 2nd instar for the tachinid, the parasite larvae matured to adults. Otherwise, *B. thuringiensis* showed no adverse effects on the parasites. Rabb (1957) noted that predatory wasps which prey on hornworms were not affected by *B. thuringiensis*. According to Tamashiro (1960), the larvae of *Corcyra cephalonica* paralyzed by the venom of *Bracon* sp. were more susceptible to spores of the bacillus innoculated into the hemocoel than unparalyzed larvae.

Insect pathologists in the Soviet Union were among the first to test mixtures of pathogens and chemical insecticides against insect pests (Telenga, 1958). Their earliest attempts were made with fungi using reduced concentrations of chlorinated hydrocarbons, such as DDT and benzene hexachloride; later the tests included *B. thuringiensis*. Successful insect control with little, if any, environmental pollution was noted. Other workers, encouraged by the successes in the Soviet Union, tested similar mixtures, but they reported conflicting results (Genung, 1960; McEwen et al., 1960; Creighton et al., 1961).

Early field tests with *B. thuringiensis* revealed that it could be used as sprays and dusts with various equipment, including aircraft that were used for spraying of chemical insecticides. Although there were some problems in the early formulations that resulted in plant damage, or poor coverage and adherence to the foliage, *B. thuringiensis* was found compatible with most of the common adjuvants used with chemical insecticides.

III. PRESENT ACCOMPLISHMENTS

By 1960 most of the favorable reports of *B. thuringiensis* in integrated control were based largely on qualitative studies; nonetheless, the early tests laid the foundation for quatitative studies over the next two decades.

Proof of the safety of *B. thuringiensis* for vertebrates and nontarget organisms now has been well documented (Krieg, 1978). The U. S. Government has approved the use of *B. thuringiensis* for numerous crop plants. The safety of the bacillus, however, depends on the selective use of varieties which do not produce toxins that adversely affect vetebrates, particularly the *B. thuringiensis* β-exotoxin or fly factor. This toxin inhibits nucleotidases and DNA dependent RNA polymerase involved in ATP synthesis, and causes teratologies in susceptible organisms. The most widely used *B. thuringiensis* commercial preparation is formulated with the *kurstaki* (HD-1) strain which does not produce the β-exotoxin.

The most effective toxin, especially for lepidopterous insects, is the δ-endotoxin contained in the parasporal crystal. Substantial testing for safety has shown that it is not toxic to vertebrates. In fact, reports from India indicate that the δ-endotoxin inhibits the growth of tumor cells in mice. For example, Prasad and Shethna (1976) and Rao et al. (1979) have observed that the δ-endotoxin, when applied at sites close to tumor cells, is cytotoxic to the tumor cells but not to normal cells. If this effect can be further substantiated, *B. thuringiensis* may have medical application. Nishiitsutsuji-Uwo et al. (1980), however, have not been able to confirm the cytotoxic action with malignant vertebrate cells *in vitro*. In fact, both normal and malignant vertebrate cells were unaffected by the δ-endotoxin, although insect cells were destroyed.

Numerous reports have appeared on the safety of *B. thuringiensis* for parasites and predators. The observations made by Biliotti (1956a) on the reaction of braconid parasites in host larvae killed by *B. thuringiensis* was confirmed by Marchal-Segault (1975a,b). Marchal-Segault also noticed that parasitized host larvae were more susceptible to the bacillus than unparasitized larvae and that the susceptibility increased with the stage of development and the number of parasite larvae in the host.

Recently, several reports have indicated that at high dosages *B. thuringiensis* caused infection of parasites and predators under laboratory conditions (Marchal-Segault, 1974; Kiselek, 1975; Hamed, 1979). In general, larvae were more susceptible than adults. Table 1 shows the infectivity of *B. thuringiensis* for parasites and predators of *Yponomeuta evonymellus*. Hassan and Krieg (1975) reported that *B. thuringiensis* does not infect the adult of *Trichogramma cacoecia*.

The presence of β-exotoxin in preparations of *B. thuringiensis* has caused adverse effects on insect parasites and predators. For example, Kiselek (1975) reported that the presence of the β-exotoxin caused death or teratologies to larvae of *Chrysopa carnea* and *Coccinella septempunctata*, but not to larvae of *Trichogramma pallida*. The preparation, Bitoxibacillin-202, which contains spores, δ-endotoxin, and β-exotoxin, caused mortality of

TABLE 1. Infectivity of *Bacillus thuringiensis* for parasites and predators of *Yponomeuta evonymellus*[a].

SPECIES	FAMILY	% PARASITISM OF *Yponomeuta*	BACTERIA IN HEMOLYMPH[b]
Diadegma armillata	Ichneumonidae	3.11	+
Pimpla turionellae	Ichneumonidae	0.95	+
Trichionotus sp.	Ichneumonidae	0.37	-
Ageniaspis fuscicollis	Encyrtidae	1.87	+
Tetrastichus evonymellae	Eulophidae	1.23	+
Bessa fugax	Tachinidae	0.30	-
Zenillia dolosa	Tachinidae		-
Picromerus bidens	Hemiptera	-	-

a. Modified after Hamed (1979).
b. + = infected and died of septicemia.

some of the parasites and predators associated with the codling moth (Kazakova and Dzhunusov, 1975).

Several years ago, Forsberg et al. (1976) authored a controversial monograph directed specifically to issues raised about the use of *B. thuringiensis* for spruce budworm control in Canada. The report indicated that because some of the published experimentation was either inadequate or faulty, the safety of *B. thuringiensis* had not been substantially documented. I do not agree with this conclusion but shall leave the rebuttal to those individuals who have more expertise in this area. Nonetheless, I wish to emphasize that some of the questions and problems raised by Forsberg et al. (1976) should not be ignored. These questions may equally apply to other pathogens used for insect control.

Studies that integrate insect parasites with *B. thuringiensis* preparations suggest that the combination is more effective than when applied separately against such insects as the gypsy moth, *Porthetria dispar* (Wollam and Yendol, 1976; Ahmad et al., 1978), the codling moth, *Laspeyresia pomonella* (Karadzhov, 1974), the cutworm, *Agrotis segetum* (Kashkarova, 1975), and *Heliothis* spp. (Bull et al., 1979). With the cutworm, sublethal doses of entobakterin (var. *galleriae*) or dendrobacillin (var. *dendrolimus*) applied with *Apanteles* spp. increased cutworm mortality by 10-20% when compared to *Apanteles* spp. applied alone. At sublethal doses, the parasites developed normally. Aerial application of *B. thuringiensis* and Sevin on a dense population of gypsy moth drastically reduced the gypsy moth population and provided excellent protection of the foliage (Reardon et al., 1979). The parasite population also decreased, apparently owing to a reduction in the host population rather than through direct mortality of the adult parasites from the application of the mixture.

Combinations of *B. thuringiensis* with other insect pathogens in control programs have resulted either in complementation or in antagonism. Some mixtures of different varieties of *B. thuringiensis* are synergistic (Zurabova, 1968), and others are antagonistic (Pendleton, 1969). The antagonism apparently is caused by an antibiotic, bacteriocin (thuricin), described by Krieg (1970). Twelve varieties of *B. thuringiensis* have been described as bacteriocin producers (de Barjac and Lajudie, 1974).

Following the suggestions of Steinhaus (1956a,b) and Thompson (1958), attempts have been made to use *B. thuringiensis* with insect viruses. These mixtures have given favorable control of the great basin tent caterpillar, *Malacosoma fragile* (Stelzer, 1965; 1967), the Douglas fir tussock moth, *Orgyia pseudotsugata* (Stelzer et al., 1975), the tobacco budworm, *Heliothis virescens* (Bell and Romine, 1980), and a number of lepidopterous larvae on cabbage (Creighton et al., 1970). However, instances of no effect or even reduced effectiveness with *B. thuringiensis*-virus mixtures have been reported. For example Young et al. (1980) reported that a mixture of *B. thuringiensis* and the nuclear polyhedrosis virus of the cabbage looper, *Trichoplusia ni*, at dosages causing an LC45 or greater resulted in increased larval mortality; at lower dosages antagonism between the two pathogens resulted in decreased larval mortality. In fact, larval mortality was greater and more rapid when the bacillus treatment followed the virus exposure than when both pathogens were applied simultaneously. These workers concluded that the bacillus-virus mixture was of no advantage in controlling the cabbage looper. On the other hand, McVay et al. (1977) reported an additive response with a similar bacillus-virus mixture used against the cabbage looper.

A mixture of *B. thuringiensis* and a cytoplasmic polyhedrosis virus caused higher mortality of larvae of the pine caterpillar, *Dendrolimus spectabilis*, than either pathogen by itself (Katagiri and Iwata, 1976; Katagiri et al., 1977). The virus was the major mortality factor in natural populations of the pine caterpillar, and its application caused larval death in about 3 weeks. The addition of *P. thuringiensis* produced faster mortality than with the virus used alone and also reduced larval feeding. According to Katagiri et al. (1977), the percentage of surviving adults was lower in plots treated with both pathogens than in those treated with each pathogen separately (Table 2).

B. thuringiensis and the microsporidan *Nosema pyraustae* have been combined in attempts to control the European corn borer (Lublinkhof et al., 1979; Lublinkhof and Lewis, 1980). The larval mortalities produced by the mixed preparation were additive. Lublinkhof et al. (1979) concluded that if the microsporidan were introduced and established in an insect population, the additional use of *B. thuringiensis* or chemical insecticides could provide a very effective pest management strategy for suppressing the European corn borer.

Proper formulations, as with chemical insecticides, are essential for the effective application of *B. thuringiensis*, especially where extensive or specific coverage of host plants is required (e.g., forests). For example, effective control of the European corn borer requires granular or foam formulations of *B. thuringiensis* for optimum persistance and efficacy as compared to spray formulations (Lynch et al., 1980).

Most adjuvants used with chemical insecticides have been tested with preparations of *B. thuringiensis* and have been found to be compatible (Herfs, 1965; Angus and Lüthy, 1971; Falcon, 1971). Angus and Lüthy (1971) listed 45 adjuvants that have been

TABLE 2. Effectiveness of *Bacillus thuringiensis* and cytoplasmic polyhedrosis virus for the control of *Dendrolimus spectabilis*[a].

TREATMENT	% ADULT EMERGENCE
B.t.	5.1
CPV	2.8
B.t. + CPV[b]	0.8
Control	4.8

[a]Katagiri et al. (1977).

[b]CPV at 10% lower dose than for CPV alone.

tested with preparations of *B. thuringiensis*, and the number undoubtedly has increased in the past 10 years. These adjuvants improve the wetting, spreading, and sticking of spray formulations to foliage, or serve as carriers for dust preparations, or as feeding stimulants and attractants for target insects. In addition, some adjuvants serve as antidegradients or protectants, especially in sunlight, to prolong the activity of spores and toxins. Krieg (1975) reported that some of the more effective UV protectants were fluochromes, dipicolinic acid, riboflavin, RNA, and India ink.

Most of the additives have no significant effect on the viability of spores of *B. thuringiensis* (Table 3). The few additives reported incompatible with the bacillus are the emulsifiers, Atlox and Triton X-100 (Morris, 1975a). In some cases, when added to the bacillus preparation, the adjuvants do not enhance insect control. Schuster (1979) tested various wetting agents, stickers, antidegradients, attractants, and feeding stimulants and concluded that these adjuvants are ineffective for use in *B. thuringiensis* pest management systems directed at control of the cabbage looper on cabbage.

Two other adjuvants, boric acid and chitinase, are worth mentioning. Doane and Wallace (1964) first reported that boric acid at 1% concentration markedly increased the mortality of gypsy moth when mixed with preparations of *B. thuringiensis*. Boric acid by itself had no effect. Field studies also have demonstrated an enhanced effect in the control of *Spodoptera litura* with mixtures of boric acid and *B. thuringiensis* (Govindarajan et al., 1976).

TABLE 3. Effect of sticker-spreaders on the viability of commercially produced *Bacillus thuringiensis* spores[a].

ADDITIVE	VIABLE SPORE PER g[b] THURICIDE	ENTOBACTERINE
Control (no additive)	7.6×10^9	5.9×10^7
Lovo	7.6×10^9	7.0×10^7
Later's	34.0×10^9	5.2×10^7
Igepal	26.0×10^9	8.4×10^7
Triton	7.3×10^9	4.5×10^7
Hi-Spred	2.7×10^9	4.5×10^7
X-77	4.4×10^9	5.6×10^7
Plyac	6.9×10^9	3.4×10^7
Pinolene	5.0×10^9	4.2×10^7
Folicote	3.8×10^9	4.1×10^7

[a]Morris (1969)

[b]As determined by a modified pour-plate method.

Chitinase has been reported by Smirnoff (1971, 1974) to enhance the pathogenicity of *B. thuringiensis* for the spruce budworm, *Choristoneura fumiferana*. In contrast, Fast (1978) could produce no evidence of enhancement during laboratory bioassays with similar preparations. Formulations with chitinase, however, have been found to be more effective than the bacillus by itself in aerial applications against the spruce budworm (Dimond, 1972; Smirnoff, 1973; Morris, 1976; Smirnoff and Juneau, 1979). However, not all serotypes of *B. thuringiensis* produce chitinase. In a comprehensive study of 273 cultures of *B. thuringiensis* belonging to 12 serotypes, de Barjac and Casmao-Dumanoir (1975) reported great variations among serotypes in chitinase production. Serotypes 2, 4a-4b (biotype *dendrolimus*), and 5a-5c produce chitinase; serotypes 10 and 11 do not; others vary in the production of the enzyme. The question of the effectiveness of chitinase in *B. thuringiensis* preparations is still unresolved.

The very promising results with mixtures of *B. thuringiensis* and chemical insecticides reported prior to 1960 have stimulated further tests with these mixtures. The advantages of such mixtures are: (1) they enhance the pathogenicity of the bacillus, thus allowing the use of lower doses at reduced cost; (2) lower doses of chemical insecticides can be used which are safer for nontarget organisms, lowers the cost of application, and decreases pollution; (3) chemical insecticides, fungicides, and miticides can be included in formulations for pests and plant diseases not affected by the bacillus alone and this practice in some instances has reduced the cost of application. The number of mixtures tested has increased substantially during the past two decades and results have been reported by a number of investigators (Fedorintchik, 1964; Herfs, 1965; Pristavko, 1967a,b; Angus and Lüthy, 1971; Falcon, 1971). By 1971, Angus and Lüthy had listed 52 chemicals, of which 28 were insecticides and 20 were fungicides, that had been used with *B. thuringiensis* preparations successfully.

As with mixtures of *B. thuringiensis* and other insect pathogens, both complementation and antagonism have resulted from combinations of *B. thuringiensis* and chemical insecticides. Benz (1971), in a thorough discussion of the interaction between microorganisms and chemical insecticides, pointed out that synergism and antagonism were related more to the concentration of each component in the mixture than to the microbial species used. This relationship was described by Haenggi (1965) in tests with *B. thuringiensis* and DDT (Table 4). At low concentrations of DDT, synergism occurred, but at high concentrations, DDT inhibited the effect of the bacillus. In general, the synergistic effect was absent when high doses of the bacterial preparation was used; when low doses of the bacillus and low or sublethal doses of the chemical insecticide were applied, a synergistic effect occurred (Pristavko, 1967b; Morris, 1975b; Canivet et al., 1978). In addition to concentration, the type of interaction depended on larval age, insect species, and the time of year when the mixture was applied. The use of persistent and broad spectrum insecticides, even at low dosages, is questionable. The persistent chemical insecticides may still pollute the environment as they accumulate from repeated applications.

Some of the antagonistic action of chemical insecticides is associated with their poisonous effect on the vegetative growth, sporulation, or germination of *B. thuringiensis*. Several laboratory tests have been conducted with cultures exposed to various insecticides (Chen et al., 1974; Morris, 1977a). Morris (1977a) tested 27 chemical insecticides representing the organophosphates, carbamates, pyrethrins, and chlordimeform on the formation and

TABLE 4. Interaction between *Bacillus thuringiensis* and DDT on the 4th-instar larvae of *Agrotis ypsilon*[a].

TREATMENT	DOSAGE (%)	% MORTALITY
B.t.	0.1	3
	0.05	0
	0.025	0
DDT (75%)	0.05	63
	0.025	53
	0.0125	0
B.t. + DDT (75%)	0.05	90
	0.025	53
	0.0125	57

[a]Modified after Haenggi (1965).

[b]*B. thuringiensis* at a constant concentration of 0.1%.

germination of bacterial spores. Listed as having high compatibility were Orthene, Dylox, Lannate, Sevin, Zectran, and Dimilin. In general the carbamates were the most suitable for mixing with *B. thuringiensis*.

Several studies in which increased insect mortality was obtained with combinations of *B. thuringiensis* and chemical insecticides have been reported. A mixture of Dipel (*B. thuringiensis* var. *kurstaki*) and Orthene (acephate-organophosphorus) resulted in long term effective control of the spruce budworm, *Choristoneura fumiferana* (Morris, 1977b). Morris (1977b) concluded that combinations of *B. thuringiensis* and chemical pesticides was a viable alternative for spruce budworm control. Also, successful control of caterpillars on collards with a mixture of chlordimeform and *B. thuringiensis* has been reported (Creighton and McFadden, 1974). As indicated in Table 5, even applications of mixtures containing one-half the original concentration of chlordimeform gives relatively effective control.

Hamilton and Attia (1977) studied the compatibility of *B. thuringiensis* with several chemical insecticides against third-instar larvae of the diamond back moth, *Plutella xylostella* (Table 6). Of the seven chemicals tested, four enhanced and two suppressed larval mortalities. Pristavko (1967b) tested

TABLE 5. *Bacillus thuringiensis* and chlordimeform for the control of caterpillars on collards [a].

TREATMENT	DOSE/ACRE (lb)	% REDUCTION *T. ni*	% INJURED PLANTS
Dipel	0.5	78	12a[b]
Chlordimeform	0.5	87	36c
Dipel + chlordimeform	0.5ea	94	2a
Dipel	0.25	71	22b
Chlordimeform	0.25	82	25b
Dipel + chlordimeform	0.25ea	88	8a
Dipel (standard)	2.0	93	5a
Untreated	-	-	88d

[a]Modified after Creighten and McFadden (1974).

[b]Values followed by the same letters are not significantly different at the 5% level of error, based on Duncan's multiple range test.

combinations of *B. thuringiensis* and Sevin against *Cacoecia crataegina*. As indicated by Table 7, enhanced mortality occurred even at a 100-fold dilution of the bacillus suspension. In field tests with codling moth, Karadzhov (1974) used a mixture of *B. thuringiensis*, Gardona (phosphate), a fungicide, and a parasite. The results indicated that this mixture was as effective as chemical insecticides, even though Gardona was applied at 0.1 its recommended dose (Table 8). Mixtures of *B. thuringiensis* with Trichlophon (chlorophos) have been widely used in Europe for control of at least 20 different insect pests (Table 9).

Because of concerns over the environmental effects of chemical pesticides, integrated control and pest management programs are receiving greater attention. *B. thuringiensis* is a prime candidate for use in these programs. Its present use as a microbial insecticide is world wide. In the United States,

TABLE 6. Compatibility of *Bacillus thuringiensis* and certain chemical insecticides[a] tested against *Plutella xylostella*.[a]

TREATMENT	TOXICITY INDEX
B.t.	100
B.t. + binapacryl (0.01%)	554
B.t. + trichlohexyltin HCl (0.01%)	422
B.t. + chlordimeform HCl (0.01%)	422
B.t. + fentin hydroxide (0.01%)	372
B.t. + phosphamidon (0.001%)	100
B.t. + dimethoate (0.005%)	11
B.t. + demeton-S-methyl (0.05%)	23

[a]Modified after Hamilton and Attia (1977).

B. thuringiensis has been registered for use on more than 20 agricultural crops and against at least 22 insect pests (Falcon, 1971). Most of the more extensive studies on the use of *B. thuringiensis* in integrated control have been conducted in Europe, particularly in the Soviet Union. China also is actively using the bacillus in pest management programs (Hussey, 1978; Way, 1978).

Recently, a new strain (*B. thuringiensis* var. *israelensis*) was found to be highly effective against several aquatic insect larvae. Goldberg and Margalit (1977) isolated and reported its high virulence and pathogenicity for mosquitoes. Mosquitoes of the genera *Aedes, Culex, Culiseta, Psorophora*, and *Uranotaenia* are highly susceptible, but *Anopheles* is less susceptible (Goldberg and Margalit, 1977; de Barjac, 1978b; Tyrell et al., 1979; Garcia and Des Rochers, 1979; de Barjac and Coz, 1979; Lüthy et al., 1980; Hembree et al., 1980; Garcia et al., 1980b). The mortality caused by *B. thuringiensis* var. *israelensis* is so rapid that the δ-endotoxin appears to be the major cause of larval death. The symptomology is similar to that described for lepidopterous larvae (de Barjac, 1978a, 1978c; Tyrell et al., 1979). In addition to mosquitoes, *B. thuringiensis* var. *israelensis* is pathogenic for members of

TABLE 7. Effect of Entobacterin and Sevin on *Cacoecia crataegana*[a].

DOSAGE		NO. DEAD		
ENTOBACTERIN (NO. SPORES)	SEVIN (G. ACTIVE)	LARVAE[b] BY FIFTH DAY	% MORTALITY	LD_{50}
10^6	0	14	70	5.5×10^5
10^5	0	6	30	5.5×10^5
10^4	0	3	15	5.5×10^5
0	5×10^{-7}	1	5	
10^6	5×10^{-7}	19	95	2.5×10^4
10^5	5×10^{-7}	15	75	2.5×10^4
10^4	5×10^{-7}	8	40	2.5×10^4
0	0	0	0	

[a]Modified after Pristavko (1967a)

[b]Twenty larvae per treatment.

TABLE 8. Mixtures of *Bacillus thuringiensis*, fungicide, and parasite for the control of the codling moth[a].

TREATMENT[b]	% WORMINESS 1970	1971
Phosthiol (0.12%) + Gardona (0.08%)	1.45	3.30
Entobacterin (0.6%) + Phosthiol (0.012%) + Gardona (0.008%)	6.45	4.46
Entobacterin (0.6%) + Gardona (0.008%) + *Trichogramma*	3.54	2.31
Control	54.03	47.10

[a]Modified after Karadzhov (1974)

[b]All treatments with fungicide; *B. thuringiensis* + 0.1 standard concentration of insecticides.

Simuliidae, Dixidae, Ceratopogonidae, and Chironomidae (Undeen and Nagel, 1978; Garcia et al., 1980a). Garcia et al. (1980a) have tested *B. thuringiensis* var. *israelensis* against a large number of other aquatic organisms and have found that only members of the order Diptera are susceptible. (Table 10).

B. thuringiensis var. *israelensis* is a very promising candidate for control of aquatic insects that are vectors of diseases affecting humans and domestic animals (Garcia and Huffaker, 1979). The World Health Organization (WHO) is now undertaking massive field trials with the bacillus for the control of mosquito vectors of malaria and filariasis and the black fly, *Simulium damnosum*, a vector of onchocerciasis (Collins, 1979).

TABLE 9. Pests controlled by a mixture of *Bacillus thuringiensis* and Trichlorphon (chlorophos).

PEST INSECT	REFERENCE
Hyphantria cunea	Melanikova & Pyshkalo (1976)
Eutromula pariana	Tereshchenko (1976)
Adelphocoris lineolatus	Yunusov (1974)
Tortrix viridana	Znamenskii (1975)
Lymantria dispar	Khamdam-Zade (1966)
Laspeyresia pomonella	Boldyrev (1977)
Hyponomeuta malinellus	Borisoglebskaya (1977)
Loxostege sticticalis	Dyadechko et al. (1976)
Dendrolimus sibiricus	Mashanov et al. (1976)
Pieris rapae	Tsyan et al. (1963)
Sheep botfly (Oestridae)	Sartbaev (1977)
Spodoptera littoralis	Altahtawy & Abaless (1973)
Orchard pests (e.g., *Caliroa cerasi*, *Aporia crataegi*, *Nygmia phaeorrhoea*, *Abraxas grossulariata*, *Incurvallia capitella*, etc.)	Golovshchikov (1976)
Cabbage pests (e.g., *Pieris brassicae*, *Pieris rapae*, *Plutella xylostella*, etc.)	Voltukohv (1975)

TABLE 10. Infectivity of *Bacillus thuringiensis* var. *israelensis* for aquatic organisms[a].

NONSUSCEPTIBLE	SUSCEPTIBLE
Amphibia (3)[b]	Diptera:
Pisces (3)	Dixidae (1)
Amphipoda (3)	Ceratopogonidae (1)
Decapoda (1)	Chironomidae (2+)
Anostraca (1)	
Cladocera (1)	
Ostracoda (2)	
Copepoda (1)	
Isopoda (1)	
Tubularia (1)	Target organisms:
Gastropoda (1)	Culicidae
Pelecypoda (1)	Simuliidae
Insecta:	
Ephemeroptera (1)	
Odonata (2)	
Hemiptera (8)	
Coleoptera (4)	
Trichoptera (1)	
Diptera (Ephydridae) (1)	

[a]Modified after Garcia et al. (1980a).

[b]Number in parenthesis indicates number of species tested.

IV. FUTURE GOALS AND CONCLUSIONS

Past accomplishments strongly indicate a bright future for *B. thuringiensis* in integrated control and pest management programs. There are still problems and questions, however, associated with formulation, particularly for aerial application. More effective and efficient formulations are needed. These formulations should be compatible with adjuvants, chemical insecticides, other pathogens, insect parasites, and predators.

There is also a need for the selection and development of more virulent strains for use against *Heliothis* spp., the codling moth, armyworms, anopheline mosquitoes, and other insects difficult to control with present strains. Problems of spore pollution in food products, or where spores infect nontarget insects need resolving. Nonsporeforming mutants have been obtained by Nishiitsutsuji-Uwo and Wakisaka (1975) and by Johnson et al. (1980) with mutagens. The δ-endotoxin of *B. thuringiensis* produced by these mutants apparently is as effective as the toxin from the normal wild strains of *B. thuringiensis*. Such strains should be further developed and tested.

Strains also should be developed and tested for use in long term control of insect pests. To date, only a few insect populations in nature, such as the Siberian silkworm, *Dendrolimus sibiricus* (Talalaev, 1958, 1959; Gukasyan, 1961), and the lepidopteran forest pest, *Selenephera lunigera* (Talalayeva, 1967), have been noted to succumb to epizootics produced by *B. thuringiensis*.

In 1980 the U.S. Supreme Court passed judgment that life forms produced by genetic engineering technology can be patented (Wade, 1980). This ruling has spurred industry to engineer new bacterial forms for specific purposes through transformation, transduction, or conjugation. Transduction in *B. thuringiensis* has been achieved with bacteriophages (Thorne, 1978; Perlak et al., 1979). Perlak et al. (1979) used a bacteriophage, TP-13, to convert a oligosporogenic acrystalliferous mutant to a form which produces spores and crystals at high frequencies. Plasmids have been described in *B. thuringiensis* (Ermakova et al., 1978; Faust et al., 1979; Gonzàlez and Carlton, 1980; Iizuka et al., 1981a,b; Miteva, 1978; Stahly et al., 1978a; Zakharyan et al., 1979) and there is evidence that plasmid DNA may be involved in the synthesis of the parasporal crystals (Debabov et al., 1977; Ermakova et al., 1978; Galuska and Aziabekyan, 1977; Stahly et al., 1978b). Martin et al., (1981) and Alikhanian et al., (1981), have described transformation of *B thuringiensis* protoplasts with foreign plasmids DNAs. We are in the exciting era of creating new bacterial forms for use in resolving specific industrial problems and *B thuringiensis* is a worthy candidate for such engineering. Care should be taken, however, to ensure that such newly created forms are safe for nontarget organisms.

V. REFERENCES

Ahmad, S., O'Neill, J. R., Mague, D. L., and Nowalk, R. K. (1978). Toxicity of *Bacillus thuringiensis* to gypsy moth larvae parasitized by *Apanteles melanoscelus*. *Environ. Entomol.*, 7, 73-76.

Alikhanian, S. I., Ryabchenko, N. F., Bukanov, N. O., and Sakanyan, V. A. (1981). Transformation of *Bacillus thuringiensis* subsp. *galleria* protoplasts by plasmid pBC16. *J. Bacteriol.*, 146, 7-9.

Altahtawy, M. M. and Abaless, I. M. (1973) An integrated control trial of *Spodoptera littoralis* (Boisd).) using *Bacillus thuringiensis* associated with insecticides. *Z. Angew. Entomol.*, 74, 255-263.

Angus, T. A. and Lüthy, P. (1971). Formulation of microbial insecticides. *In* "Microbial Control of Insects and Mites" (H. D. Burges and N. W. Hussey, eds.), pp. 623-638. Academic Press, New York.

de Barjac, H. and Lajudie, J. (1974). Mise en évidence de facteurs antagonistes du type des bactériocines chez *Bacillus thuringiensis*. *Ann. Microbiol.*, 125B, 529-537.

de Barjac, H. and Cosmao-Dumanoir, V. (1975). Répartition des pouvoirs chitinolytiques et lipolytiques chez les divers sérotypes de *Bacillus thuringiensis*. *Entomophaga*, 20, 43-48.

de Barjac, H. (1978a). Etude cytologique de l'action de *Bacillus thuringiensis* var. *israelensis* sur larves de Moustiques. *C. R. Acad. Sci. Paris, Ser. D.*, 286, 1629-1632.

de Barjac, H. (1978b). Un nouveau candidat à la lutte biologique contre les moustiques: *Bacillus thuringiensis* var. *israelensis*. *Entomophaga*, 23, 309-319.

de Barjac, H. (1978c). Toxicité de *Bacillus thuringiensis* var. *israelensis* pour les larves d'*Aedes aegyptii* et d'*Anopheles stephensi*. *C. R. Acad. Sci. Paris, Ser. D.*, 286, 1175-1178.

de Barjac, H. and Coz, J. (1979). Sensibilité comparée de six espèces différentes de moustiques à *Bacillus thuringiensis* var. *israelensis*. Bull. Organ. Mondiale Santé, 57, 139-141.

Bell, M. R. and Romine, C. L. (1980). Tobacco budworm field evaluation of microbial control in cotton using *Bacillus thuringiensis* and a nuclear polyhedrosis virus with a feeding adjuvant. *J. Econ. Entomol.*, 73, 427-430.

Benz, G. (1971). Synergism of micro-organisms and chemical insecticides. *In* "Microbial Control of Insects and Mites" (H. D. Burges and N. W. Hussey, eds.), pp. 327-355. Academic Press, New York.

Berliner, E. (1915). Über die Schlaffsucht der Mehlmottenraupe (*Ephestia kuhniella* Zell.) und ihren Erreger *Bacillus thuringiensis* n.sp. *Z. Angew. Entomol.*, 2, 29-56.

Biliotti, E. (1956a). Entomophages et maladies des insectes. *Entomophaga*, 1, 45-53.

Biliotti, E. (1956b). Relations entre agents pathogènes et entomophages. *Entomophaga*, 1, 101-103.

Boldyrev, M. I. (1977). Against the codling moth. *Zashch. Rast.*, 5, 24.

Borisoglebskaya, M. S. (1977). The apple moth. *Zashch. Rast.*, 5, 60-61.

Bull, D. L., House, V. S., Ables, J. R., and Morrison, R. K. (1979). Selective methods for managing insect pests of cotton. *J. Econ. Entomol.*, 72, 841-846.

Canivet, J. P., Nef, L., and Lebrun, Ph. (1978). Utilisation combinée de *Bacillus thuringiensis* et d' insecticides chimiques à does réduites contre *Euproctis chrysorrhoea*. *Z. Angew. Entomol.*, 86, 85-97.

Chen, Ker-Sang, Funke, B. R., Schultz, J. T., Carlson, R. B., and Proshold, F. I. (1974). Effects of certain organophosphate and carbamate insecticides in *Bacillus thuringiensis*. *J. Econ. Entomol.*, 67, 471-473.

Chorine, V. (1930). On the use of bacteria in the fight against the corn borer. *Int. Corn Borer Invest.*, *Sci. Rep.*, 3, 94-98.

Collins, P. (1979). WHO looks at bacteria for vector control. *Nature*, (London), 282, 349.

Creighton, C. S., Kinard, W. S., and Allen, N. (1961). Effectiveness of *Bacillus thuringiensis* and several chemical insecticides for control of budworms and hornworms on tobacco. *J. Econ. Entomol.*, 54, 1112-1114.

Creighton, C. S., McFadden, T. L., and Bell, J. V. (1970). Pathogens and chemicals tested against caterpillars on cabbage. *USDA, ARS, Prod. Res. Rep.*, No. 114. 10 pp.

Creighton, C. S. and McFadden, T. L. (1974). Complementary actions of low rates of *Bacillus thuringiensis* and chlordimeform hydrochloride for control of caterpillars on cole crops. *J. Econ. Entomol.*, 67, 102-104.

Debabov, V. G., Azizbekyan, R. R., Chilabalina, O. I., Djarhenko, V. V., Galushka, F. P., and Belich, R. A. (1977). Isolation and preliminary characterization of extrachromosomal elements of *Bacillus thuringiensis* DNA. *Genetica*, (USSR). 13, 496-500.

Dimond, J. B. (1972). A demonstration of *Bacillus thuringiensis*, plus the enzyme chitinase, against the spruce budworm in Maine. Part I. Efficacy. *Misc. Rep. Maine Agric. Exp. Sta.*, 144. 26 pp.

Doane, C. C. and Wallis, R. C. (1964). Enhancement of the action of *Bacillus thuringiensis* var. *thuringiensis* Berliner on *Porthetria dispar* (Linnaeus) in laboratory tests. *J. Insect Pathol.*, 6, 423-429.

Dyadechko, N. P., Tsybul'skaya, G. N., Chizhik, R. I., and Venger, V. M. (1976). Biological agents reducing the numbers of the meadow moth. *Zashch. Rast.*, 7, 43-44.

Ellinger, T. (1928). International corn borer investigations. *Int. Corn Borer Invest., Sci. Rep.*, 1927-1928. Vol. 1.

Ermakova, L. M., Galushka, F. P., Strongin, A. Ya., Sladkova, I. A., Rebentish, B. A., Andruka, M. V., and Stepanov, V. M. (1978). Plasmids of crystal-forming bacilli and the influence of growth medium composition on their appearance. *J. Gen. Microbiol.*, 107, 169-171.

Falcon, L. A. (1971). Use of bacteria for microbial control. *In* "Microbial Control of Insects and Mites" (H. D. Burges and N. W. Hussey, eds.), pp. 67-95. Academic Press, London.

Fast, P. G. (1978). Laboratory bioassays of mixtures of *Bacillus thuringiensis* and chitinase. *Can. Entomol.*, 110, 201-203.

Faust, R. M., Spizizen, J., Gage, V., and Travers, R. S. (1979). Extrachromosomal DNA in *Bacillus thuringiensis* var. *kurstaki*, var. *finitimus*, var. *sotto*, and in *Bacillus popilliae*. *J. Invertebr. Pathol.*, 33, 233-238.

Fedorintchik, N. S. (1964). Les facteurs déterminant l'efficacité des biopréparations dans la protection des végétaux. *Entomophaga, Mem. Hors Ser. No.* 2, pp. 51-61.

Fisher, R. A. and Rosner, L. (1959). Toxicology of the microbial insecticide, Thuricide. *Agric. Food Chem.*, 7, 686-688.

Forsberg, C. W., Henderson, M., Henry, E., and Roberts, J. R. (1976). *Bacillus thuringiensis*: its effects on environmental quality. *Natl. Res. Coun. Can. No.* 15385. 134 pp.

Galushka, F. P. and Azizbekyan, R. R. (1977). Investigations of plasmids of strains of different variants of *Bacillus thuringiensis*. *Dokl. Akad. Nauk. SSR.*, 236, 1233-1235.

Garcia, R. and Huffaker, C. B. (1979). Ecosystem management for suppression of vectors of human malaria and schistosomiasis. *Agro-Ecosystems*, 5, 295-315.

Garcia, R. and Des Rochers, B. (1979). Toxicity of *Bacillus thuringiensis* var. *israelensis* to some California mosquitoes under different conditions. *Mosq. News*, 39, 541-544.

Garcia, R., Des Rochers, B., and Tozer, W. (1980a). Studies on the toxicity of *Bacillus thuringiensis* var. *israelensis* against organisms found in association with mosquito larvae. *Proc. Calif. Mosq. Vector Assoc.*, 48, 33-36.

Garcia, R., Federici, B., Hall, J. M., Mulla, M. S., and Schaefer, C. H. (1980b). Bti - a potent new biological weapon. *Calif. Agric.*, 34, 18-19.

Genung, W. G. (1960). Comparison of insecticides, insect pathogens and insecticide-pathogen combinations for control of cabbage looper *Trichoplusia ni* (Hbn.). *Fl. Entomol.*, 43, 65-68.

Goldberg, L. J. and Margalit, J. (1977). A bacterial spore demonstrating rapid larvicidal activity against *Anopheles sergentii, Uranotaenia unguiculata, Culex univittatus, Aedes aegypti* and *Culex pipiens*. *Mosq. News*, 37, 355-358.

Golovshchikov, N. G. (1976). Entobakterin -- an effective preparation. *Zashch. Rast.*, 6, 62.

González, J. M., Jr. and Carlton, B. C. (1980). Patterns of plasmid DNA in crystalliferous and acrystalliferous strains of *Bacillus thuringiensis*. *Plasmid*, 3, 92-98.

Govindarajan, R., Jayaraj, S., and Narayanan, K. (1976). Mortality of the tobacco caterpillar, *Spodoptera litura* (F.), when treated with *Bacillus thuringiensis* combinations with boric acid and insecticides. *Phytoparasitica*, 4, 193-196.

Gukasyan, A. B. (1961) Prospective control method for the Siberian silkworm. *Akad. Nauk S.S.S.R. Vestn.*, 31, 58-59.

Haenggi, A. (1965). L'efficacité d'un mélange de *Bacillus thuringiensis* Berl. et de DDT contre *Agrotis ypsilon* Rott. (Noctuidae). *Entomophaga*, 10, 343-348.

Hamed, A. R. (1979). Zur Wirkung von *Bacillus thuringiensis* auf Parasiten ünd Pradatoren von *Yponomeuta evonymellus* (Lep., Yponomeutidae). *Z. Angew. Entomol.*, 87, 294-311.

Hamilton, J.T. and Attica, F. I. (1977). Effects of mixtures of *Bacillus thuringiensis* and pesticides on *Plutella xylostella* and the parasite *Thyraeella collaris*. *J. Econ. Entomol.*, 70, 146-148.

Hassan, S. and Krieg, A. (1975). Über die schonende Wirkung von *Bacillus thuringiensis* - Präparaten auf den Parasiten *Trichogramma cacoeciae* (Hym.: Trichogrammatidae). *Z. Pflanzenkr. Pflanzenpathol. Pflanzenschutz*, 82, 515-521.

Hembree, S. C., Meisch, M. V., and Williams, D. (1980). Field test of *Bacillus thuringiensis* var. *israelensis* against *Psorophora columbiae* larvae in small rice plots. *Mosq. News*, 40, 67-70.

Herfs, W. (1965). Die Verträglichkeit von *Bacillus thuringiensis* - Präparaten mit chemischen Pflanzenschutzmitteln und mit Beistoffen. *Z. Pflanzenkr. Pflanzenpathol. Pflanzenschutz*, 72, 584-599.

Hussey, N. W. (1978). Biological control techniques. *The Royal Society Delegation on Biological Control to China. The Royal Society*, London., pp. 7-14.

Husz, B. (1928). *Bacillus thuringiensis* Berl., a bacterium pathogenic to corn borer larvae. A preliminary report. *Int. Corn Borer Invest., Sci. Rep.* 1927-1928., 1, 191-193.

Husz, B. (1929). On the use of *Bacillus thuringiensis* in the fight against the corn borer. *Int. Corn Borer Invest., Sci. Rep.*, 2, 99-105.

Iizuka, T., Faust, R. M., and Travers, R. S. (1981a). Isolation and partial characterization of extrachromosomal DNA from serotypes of *Bacillus thuringiensis* pathogenic to lepidopteran and dipteran larvae by agarose gel electrophoresis. *J. Sericult. Sci. Japan.*, 50, 1-14.

Iizuka, T., Faust, R. M., and Travers, R. S. (1981b). Comparative profiles of extrachromosomal DNA in single and multiple crystalliferous strains of *Bacillus thuringiensis* var. *kurstaki*. *J. Fac. Agr. Hokkaido Univ. Japan.*, 60, 143-151.

Ishiwata, S. (1901). On a severe flacherie (sotto disease). *Dainihon Sanshi Kaiho*, No. 114, 1-5.

Johnson, D. E., Niezgodski, D. M., and Twaddle, G. M. (1980). Parasporal crystals produced by oligosporogenous mutants of *Bacillus thuringiensis* (Spo^- Cry^+). *Can. J. Microbiol.*, 26, 486-491.

Karadzhov, S. (1974). Effectiveness of entobacterin (*Bacillus thuringiensis* var. *galleriae*) against the apple codling moth (*Carpocapsa pomonella* L.) *Horti. Viti. Sci.*, 11, 53-60.

Kashkarova, L. F. (1975). Simultaneous use of pathogenic microorganisms and parasitic insects against *Agrotis segetum*. *Trudy Vsesoyuznogo Nauchno-Issledovatel'Skogo Instituta Sashch. Rast.*, 42, 130-136.

Katagiri, K. and Iwata, Z. (1976). Control of *Dendrolimus spectabilis* with a mixture of cytoplasmic polyhedrosis virus and *Bacillus thuringiensis*. *Appl. Entomol. Zool.*, 11, 363-364.

Katagiri, K., Iwata, Z., Kushida, T., Fukuizumi, Y., and Ishizuka, H. (1977). Effects of application of Bt, CPV and a mixture of Bt and CPV on the survival rates in populations of the pine caterpillar, *Dendrolimus spectabilis*. *J. Japan Forest Soc.*, 59, 442-448.

Kazakova, S. B. and Dzhunusov, K. K. (1975). The effects of Bitoxi-bacillin-202 on certain orchard insects in the Issyk-kul depression. *Entomol. Invest. Kirgizia*, 10, 49-51.

Khamdam-Zade, T. K. (1966). Effectiveness of Dendrobacillin for the control of gypsy moth. *Tashkend. Vses. Nauch-Issled. Inst. Khlopkovod.*, 9, 164-166.

Kiselek, E. V. (1975). The effect of biopreparations on insect parasites and predators. *Zashch. Rast.*, 12, 23.

Krieg, A. (1970). Thuricin, a bacteriocin producted by *Bacillus thuringiensis*. *J. Invertebr. Pathol.*, 15, 291.

Krieg, A. (1975). Photoprotection against inactivation of *Bacillus thuringiensis* spores by ultraviolet rays. *J. Invertebr. Pathol.*, 25, 267-268.

Krieg, A. (1978). Insektenbekämpfung mit *Bacillus thuringiensis* - Präparaten und deren Einfluss auf die Umwelt. *Nachrichtenbl. Dtsch. Pflanzenschutzdiens.*, 30, 177-181.

Lublinkhof, J., Lewis, L. C., and Berry, E. C. (1979). Effectiveness of integrating insecticides with *Nosema pyraustae* for suppressing populations of the European corn borer. *J. Econ. Entomol.*, 72, 880-883.

Lublinkhof, J. and Lewis, L. C. (1980) Virulence of *Nosema pyraustae* to the European corn borer when used in combination with insecticides. *Environ. Entomol.*, 9, 67-71.

Lüthy, P., Raboud, G., Delucchi, V., and Küenzi, M. (1980). Field efficacy of *Bacillus thuringiensis* var. *israelensis*. *Mitt. Schweiz. Entomol. Ges.*, 53, 3-9.

Lynch, R. E., Lewis, L. C., and Berry, E. C. (1980). Application efficacy and field persistence of *Bacillus thuringiensis* when applied to corn for European corn borer control. *J. Econ. Entomol.*, 73, 4-7.

McEwen, F. L., Glass, E. H., Davis, A. C., and Splittstoesser, C. M. (1960). Field tests with *Bacillus thuringiensis* for control of four lepidopterous pests. *J. Insect Pathol.*, 2, 152-164.

McVay, J. R., Gudauskas, R. T., and Harper, J. D. (1977). Effects of *Bacillus thuringiensis* nuclear-polyhedrosis virus mixtures on *Trichoplusia ni* larvae. *J. Invertebr. Pathol.*, 29, 367-372.

Marchal-Segault, D. (1974). Sensibilité des hyménoptères braconides *Apanteles glomeratus* et *Phanerotoma flavitestacea* au complexe spores-cristaux de *Bacillus thuringiensis* Berliner. *Ann. Zool. Écol. Anim.*, 6, 521-528.

Marchal-Segault, D. (1975a). Rôle des larves entomophages dans l'infection à *Bacillus thuringiensis* Berliner des chenilles de *Pieris brassicae* L. et *Anagasta kuehniella* Zell. *Rev. Zool. Agric. Pathol. Veg.*, 74, 68-84.

Marchal-Segault, D. (1975b). Développement larvaire des Hyménoptères parasites *Apanteles glomeratus* L. et *Phanerotoma flavitestacea* F. chez des Chenilles infectées par *Bacillus thuringiensis* Berliner. *Ann. Parasitol. (Paris)*, 50, 223-232.

Martin, P. A. W., Lohr, J. R., and Dean, D. H. (1981). Transformation of *Bacillus thuringiensis* protoplasts by plasmid deoxyribonudeic acid. *J. Bacteriol.*, 145, 980-983.

Mashanov, A. I., Gukasyan, A. B., Kobzar, V. F., Oleh, S. A., and Sergeeva, V. L. (1976). Aerial application for bacterial control of the Siberian silkworm in the Tuva mountain forests. *Biologicheskikh Nauk*, 10, 90-95.

Mattes, O. (1927). Parasitäre Krankheiten der Mehlmottenlarven und Versuche über ihre Verwendbarkeit als biologische Bekämpfungsdmittel. (Zugleich ein Beitrag zue Zytologie der Bakterien.). *Ges. Beförd. Gesammel. Naturw. Sitzber. (Marburg).*, 62, 381-417.

Melanikova, R. G. and Pyshkalo, R. P. (1976). Control of the American white butterfly in the Crimea. *Zasch. Rast.*, 4, 49-501.

Metalnikov, S. and Chorine, V. (1929a). Experiments on the use of bacteria to destroy the corn borer. *Int. Corn Borer Invest., Sci. Rep.*, 2, 54-59.

Metalnikov, S. and Chorine, V. (1929b). On the infection of the gypsy moth and certain other insects with *Bacterium thuringiensis*. A preliminary report. *Int. Corn Borer Invest., Sci. Rep.*, 2, 60-61.

Metalnikov, S., Hergula, B., and Strail, D. M. (1930a). Experiments on the application of bacteria against the corn borer. *Int. Corn Borer Invest., Sci. Rep.*, 148-151.

Metalnikov, S., Hergula, B., and Strail, D. M. (1930b). Utilisation des microbes dans la lutte contre la pyrale du maïs. *C. R. Acad. Sci. Paris. Sér. D.*, 191, 738-740.

Metalnikov, S., Hergula, B., and Strail, D. M. (1931). Utilisation des microbes dans la lutte contre la pyrale du maïs. *Ann. Inst. Pasteur*, 46, 320-325.

Miteva, V. I. (1978). Isolation of plasmid DNA from various strains of *Bacillus thuringiensis* and *Bacillus cereus*. *C. R. Acad. Sci. Bulgaria*, 31, 913-916.

Morris, O. N. (1969). Susceptibility of several forest insects of British Columbia to commercially produced *Bacillus thuringiensis*. I. Studies on the physiological properties of some commercial products. *J. Invertebr. Pathol.*, 13, 134-146.

Morris, O. N. (1975a). Effect of some chemical insecticides on the germination and replication of commercial *Bacillus thuringiensis*. *J. Invertebr. Pathol.*, 26, 199-204.

Morris, O. N. (1975b). Susceptibility of the spruce budworm, *Choristoneura fumiferana*, and the white-marked tussock moth, *Orgyia leucostigmata*, to *Bacillus thuringiensis*: chemical insecticide combinations. *J. Invertebr. Pathol.*, 26, 193-198.

Morris, O. N. (1976). A 2-year study on the efficacy of *Bacillus thuringiensis*-chitinase combinations in spruce budworm (*Choristoneura fumiferana*) control. *Can. Entomol.*, 108, 225-233.

Morris, O. N. (1977a). Compatibility of 27 chemical insecticides with *Bacillus thuringiensis* var. *kurstaki*. *Can. Entomol.*, 109, 855-864.

Morris, O. N. (1977b). Long term study of the effectiveness of aerial application of *Bacillus thuringiensis*-acephate combinations against the spruce budworm, *Choristoneura fumiferana* (Lepidoptera: Tortricidae). *Can. Entomol.*, 109, 1239-1248.

Nishiitsutsuji-Uwo, J. and Wakisaka, Y. (1975). Sporeless mutants of *Bacillus thuringiensis*. *J. Invertebr. Pathol.*, 25, 355-361.

Nishiitsutsuji-Uwo, J., Endo, Y., and Himeno, M. (1980). Effects of *Bacillus thuringiensis* δ-endotoxin on insect and mammalian cells *in vitro*. *Appl. Entomol. Zool.*, 15, 133-139.

Pendleton, I. R. (1969). Ecological sugnificance of antibiotics of some varieties of *Bacillus thuringiensis*. *J. Invertebr. Pathol.*, 13, 235-240.

Perlak, F. J., Mendelsohn, C. L., and Thorne, C. B. (1979). Converting bacteriophage for sporulation and crystal formation in *Bacillus thuringiensis*. *J. Bacteriol.*, 140, 699-706.

Prasad, S. S. S. V. and Shethna, Y. I. (1976). Antitumor immunity against Yoshida ascites sarcoma after treatment with proteinaceous crystal of *Bacillus thuringiensis* var. *thuringiensis*. *Indian J. Exp. Biol.*, 14, 285-288.

Pristavko, V. P. (1967a). On the use of entomopathogenic bacteria together with insecticides in the control of insect pests. *Entomol. Rev.*, 46, 443-446.

Pristavko, V. P. (1967b). On the use of *Bacillus thuringiensis* -- insecticide combinations to control insect pests. *Proc. Int. Coll. Insect Pathol. Microbial Control. Wagenigen, The Netherlands,* Sept. 5-10, 1966. North Holland Publ. Co., Amsterdam, pp. 212-213.

Rabb, R. L. (1957). Effects of *Bacillus thuringiensis* on *Polistes exclamans. Expt. Sta., North Carolina. Inf. Note No.* 118. Mimeo Rpt., 11 pp.

Rao, A. S., Amonkar, S. V., and Phondke, G. P. (1979). Cytotoxic activity of the δ-endotoxin of *Bacillus thuringiensis* var. *thuringiensis* (Berliner) on fibrosarcoma in Swiss mice. I *J. Exp. Biol.*, 17, 1208.

Reardon, R., Metterhouse, W., and Balaam, R. (1979). Impact of aerially applied *Bacillus thuringiensis* and carbaryl on gypsy moth (Lep.: Lymantriidae) and adult parasites. *Entomophaga*, 24, 305-310.

Sartbaev, S. K. (1977). The use of entobacterin mixed with insecticides in the control of sheep botflies in the conditions of Kirgizia. *Biol. Methods Control Blood-sucking Insects Acarina: Frunze, USSR.* pp. 47-48. (*Rev. Appl. Entomol.,* 67B, 269. 1979).

Schuster, D. J. (1979). Adjuvants tank-mixed with *Bacillus thuringiensis* for control of cabbage looper larvae on cabbage. *J. G. Entomol. Soc.*, 14, 182-186.

Smirnoff, W. A. (1971). Effect of chitinase on the action of *Bacillus thuringiensis. Can. Entomol.*, 103, 1829-1831.

Smirnoff, W. A. (1973). Results of tests with *Bacillus thuringiensis* and chitinase on larvae of the spruce budworm. *J. Invertebr. Pathol.*, 21, 116-118.

Smirnoff, W. A. (1974). The symptoms of infection by *Bacillus thuringiensis* + chitinase formulation in larvae of *Choristoneura fumiferana. J. Invertebr. Pathol.*, 23, 397-399.

Smirnoff, W. A. and Juneau, A. (1979). Results of aerial spraying with *Bacillus thuringiensis* against the spruce budworm in Quebec in 1975. *Ann. Soc. Entomol. Quebec*, 24, 131-138.

Stahly, D. P., Dingman, D. W., Irgens, R. L., Field, C. C., Feiss, M. G., and Smith, G. L. (1978a). Multiple extrachromosomal deoxyribonucleic acid molecules in *Bacillus thuringiensis. FEMS Microbiol. Lett.*, 3, 139.

Stahly, D. P., Dingman, D. W., Bulla, L. A., and Aronson, A. I. (1978b). Possible origin and function of the parasporal crystals in *Bacillus thuringiensis*. *Biochem. Biophys. Res. Commun.*, 84, 581-588.

Steinhaus, E. A. (1951a). Pest control by bacteria. *Calif. Agric.*, 5, 5.

Steinhaus, E. A. (1951b). Possible use of *Bacillus thuringiensis* Berliner as an aid in the biological control of the alfalfa caterpillar. *Hilgardia*, 20, 359-381.

Steinhaus, E. A. (1956a). Potentialities for microbial control of insects. *Agric. Food Chem.*, 4, 676-680.

Steinhaus, E. A. (1956b). Living insecticides. *Sci. Amer.*, 195, 96-104.

Steinhaus, E. A. (1975). *In* "Disease in a minor chord". Ohio State Univ. Press, Columbus, Ohio.

Stelzer, M. J. (1965). Susceptibility of the great basin tent caterpillar, *Malacosoma fragile* (Stretch), to a nuclear-polyhedrosis virus and *Bacillus thuringiensis* Berliner. *J. Invertebr. Pathol.*, 7, 122-125.

Stelzer, J. J. (1967). Control of a tent caterpillar, *Malacosoma fragile incurva*, with an aerial application of a nuclear-polyhedrosis virus and *Bacillus thuringiensis*. *J. Econ. Entomol.*, 60, 38-41.

Stelzer, M. J., Neisess, J., and Thompson, C. G. (1975). Aerial applications of a nucleopolyhedrosis virus and *Bacillus thuringiensis* against the Douglas fir tussock moth. *J. Econ. Entomol.*, 68, 269-272.

Talalaev, E. V. (1958). Induction of epizootic septicemia in the caterpillars of Siberian silkworm moth, *Dendrolimus sibiricus* Tschtv. (Lepidoptera, Lasiocampidae). *Entomol. Rev.*, 37, 557-567.

Talalaev, E. V. (1959). Bacteriological method for control of Siberian silkworm. *Trans. 1st Int. Conf. Insect Pathol. Biol. Control Praha* 1958., pp. 51-57.

Talalayeva, G. B. (1967). A case of bacterial epizooty in a larval population of *Selenephera lunigera* Esp. (Lepidoptera, Lasiocampidae). *Entomol. Rev.*, 46, 190-192.

Tamashiro, M. (1960). The susceptibility of *Bracon*-paralyzed *Corcyra cephalonica* (Stainton) to *Bacillus thuringiensis* var. *thuringiensis* Berliner. *J. Insect Pathol.*, 2, 202-219.

Tanada, Y. and Reiner, C. (1960). Microbial control of the artichoke plume moth, *Platyptilia carduidactyla* (Riley) (Pterophoridae, Lepidoptera). *J. Insect Pathol.*, 2, 230-246.

Tanada, Y. and Reiner, C. (1962). The use of pathogens in the control of the corn earworm, *Heliothis zea* (Boddie). *J. Insect Pathol.*, 4, 139-154.

Telenga, N. A. (1958). Die Anwendung der Müskardinenpilze im Verein mit Insektiziden für die Bekämpfung der Schädlingsinsekten. *Trans. 1st. Int. Conf. Insect Pathol. Biol. Control, Praha,* 1958. pp. 155-168.

Tereshchenko, V. E. (1976). The apple leaf rolling moth. *Zashch. Rast.*, 5, 61.

Thompson, C. G. (1958). The use of certain entomogenous microorganisms to control the alfalfa caterpillar. *Proc. Intern. Congr. Entomol., 10th, Montreal,* 1956., 4, 693.

Thorne, C. B. (1978). Transduction in *Bacillus thuringiensis*. *Appl. Environ. Microbiol.*, 36, 1109-1115.

Tsyan, N., et al. (1963). Investigations on the use of the sporulating bacteria *Bacillus cereus* var. *galleriae* and *Bacillus thuringiensis* var. *thuringiensis*. *Acta Phytophyl. Sin.*, 2, 149-162. (*Rev. Appl. Entomol.*, 52A, 397, 1964).

Tyrell, D. J., Davidson, L. I., Bulla, L. A., Jr., and Ramoska, W. A. (1979). Toxicity of parasporal crystals of *Bacillus thuringiensis* subsp. *israelensis* to mosquitoes. *Appl. Environ. Microbiol.*, 38, 656-658.

Undeen, A. H. and Nagel, W. L. (1978). The effect of *Bacillus thuringiensis* ONR-60A strain (Goldberg) on *Simulium* larvae in the laboratory. *Mosq. News*, 38, 524-527.

Voltukhov, V. A. (1975). Economic evaluation of the chemical and biological methods. *Zashch. Rast.*, 5, 29.

Wade, N. (1980). Court says lab-made life can be patented. *Science*, (Wash. D.C.), 208, 1445.

Way, M. J. (1978). Integrated pest control. *The Royal Society Delegation on Biological Control to China. The Royal Society,* London., pp. 14-24.

Wollam, J. D. and Yendol, W. G. (1976). Evaluation of *Bacillus thuringiensis* and a parasitoid for suppression of the gypsy moth. *J. Econ. Entomol.*, 69, 113-118.

Young, S. Y., McCaul, L. A., and Yearian, W. C. (1980). Effect of *Bacillus thuringiensis* -- *Trichoplusia* nuclear polyhedrosis virus mixtures on the cabbage looper, *Trichoplusia ni*. *J. G. Entomol. Soc.*, 15, 1-8.

Yunosov, I. (1974). Biopreparations destroy the lucerne bug. *Zashch. Rast.*, 8, 32.

Zakharyan, R. A., Israelyan, Yu. A., Agabalyan, A. S., Tatevosyan, P. E., Akopyan, S. M., and Afrikyan, E. K. (1979). Plasmid DNA from *Bacillus thuringiensis*. *Microbiology*, 48, 177-180.

Znamenskii, V. S. (1975). Increase in the population of *Tortrix viridana* in the forests of the Moscow region. *Lesn. Khozi.*, 28, 88-91.

Zurabova, E. R. (1968). Use of a complex of crystalliferous bacilli as a means of increasing the efficiency of biopreparations. *XIII Int. Congr. Entomol.*, 2, 116-117.

NATURE OF PATHOGENIC PROCESS OF *Bacillus thuringiensis*

Robert M. Faust

Insect Pathology Laboratory
Agricultural Research
Science and Education Administration
U.S. Department of Agriculture
Beltsville, Maryland

I. INTRODUCTION

Bacillus thuringiensis is a rod-shaped, Gram-positive spore-forming bacterium, 1 to 1.2 μm wide and 3 to 5 μm long, which was first discovered in 1901 by the Japanese bacteriologist Ishiwata (originally named *Bacillus sotto*). Ishiwata called this organism Sotto Bacillen and noted that one of the characteristics of the infection was the speed with which the disease could spread through a silkworm colony, resulting in blackened, spore packed larval cadavers. *B. thuringiensis* resembles *B. cereus* in many of its biochemical characteristics, but is different by the presence of a crystalline-like parasporal body that it synthesizes within the cell at the time of sporulation.

The organism is known to produce several agents that are active against several orders of insects: phospholipase C (*B. thuringiensis* α-endotoxin) (Toumanoff, 1953), a thermostable exotoxin (*B. thuringiensis* β-exotoxin) (McConnell and Richards, 1959), an unidentified enzyme that may not be toxic (*B. thuringiensis* γ-exotoxin) (Heimpel, 1965), the protein parasporal crystal (*B. thuringiensis* δ-endotoxin) (Hannay, 1953), a "labile toxin" (Smirnoff and Berlinguet, 1966), a water-soluble toxin isolated from a commercial formulation (Fast, 1971), and a mouse factor exotoxin (Krieg, 1971). Krieg (1961) also has defined as toxins a bacillogenic antibiotic and a proteinase, both produced by *B. thuringiensis*. The production of these toxins varies from strain to strain and often depends on the culture conditions (Dulmage, 1970, 1971). At the present time, *B. thuringiensis* is the most important bacterium for insect regulation and is produced by several commercial companies for use on a variety of crops.

II. HISTORY

The type species of the group, *B. thuringiensis* var. *thuringiensis*, was originally isolated from diseased larvae of the Mediterranean flour moth, *Anagasta kuehniella*, by Berliner (1911, 1915), who described it and its pathogenicity. In the late 1920s and early 1930s a number of papers appeared on the effectiveness of *B. thuringiensis* var. *thuringiensis* as an agent for the biological control of the European corn borer, *Ostrinia nubilalis* (Husz, 1928, 1930, 1931; Metalnikov and Toumanoff, 1928; Metalnikov and Chorine, 1929a,b; Metalnikov, 1930; Metalnikov et al., 1930). Jacobs (1950) reported a French product, "Sporeine," to be a biological control agent for *A. kuehniella*. Steinhaus (1951) reported that in field trials sporulated cultures of *B. thuringiensis* gave encouraging results against the larvae of *Colias eurytheme*. All studies before 1951 indicated that under certain conditions, *B. thuringiensis* var. *thuringiensis* and related varieties were pathogenic to a number of lepidopteran larvae, but the mode of action of these pathogens was not determined.

Subsequently, Hannay (1953, 1956), while examining the sporulation of a number of aerobic spore-formers, noted diamond-shaped crystals free of sporangia in preparations of sporulating cultures of *B. thuringiensis* and referred to them as parasporal bodies. The crystals had been noted earlier by Berliner (1915) and Mattes (1927), but neither credited them with any function in pathogenicity. Hannay, in contrast, suggested that the crystals might be connected with the formation of a toxic substance that induced septicemia of insect larvae. Angus (1954; 1956a,b,c;

1970) later provided experimental proof of Hannay's suggestion.

During the last 25 years, numerous strains of *B. thuringiensis* with the same characteristics (parasporal body and pathogenicity for mainly lepidoptera larva) have been isolated and for a relatively long period of time many of the earlier isolated strains were considered varieties of *Bacillus cereus* (Toumanoff and LeCoroller, 1959). However, Delaporte and Beguin (1955), and then Heimpel and Angus (1958), had recommended that all the parasporal crystal-forming strains be grouped under the name of *Bacillus thuringiensis*, with the strain isolated by Berliner as the type strain. Heimpel and Angus (1958) agreed that it might be technically correct to rename these organisms varieties of *B. cereus*, but noted that the epithet *B. thuringiensis* was so entrenched in the literature that to change it would likely lead to confusion. These problems of taxonomy have now been resolved through the intensive studies of the serological and biochemical characteristics of numerous *B. thuringiensis* strains by Bonnefoi and de Barjac (1963) and de Barjac and Bonnefoi (1962, 1967, 1968). Essentially, these investigators discovered that variants within the species of *B. thuringiensis* could be differentiated serologically by comparing antibodies to their flagellar proteins (H antigens). This technique proved to be a reproducible and reliable method and is generally accepted at the present time and used in the identification of unknown strains of *B. thuringiensis*. Table 1 shows the subspecific division of *B. thuringiensis* based on H serotypes, esterase types, and production of toxins. Sixteen serotypes have been described to date.

III. EPIZOOTIOLOGY

An epizootic disease is one that affects many insects of one kind at the same time; it is comparable to an epidemic in human populations. An enzootic disease is one that has low incidence but continually affects a given insect. Enzootic and epizootic phases of a disease may oscillate depending on changes in the virulence of the pathogen(s), resistance of the host, transmissibility, population movements, density and crowding of the host, habitat, and other environmental factors.

Only during the past two decades has increased investigations of the epizootiology of insects been undertaken, and these studies have been concentrated largely on the "milky diseases" of the Japanese beetle, *Popillia japonica* (and on a few virus diseases of some insects such as the sawflies). Further, most of the information has been obtained through studies of disease outbreaks in laboratory cultures rather than in the field. For example, excess humidity induces epizootics of bacteria in the laboratory, particularly of non-sporeforming pathogens (Bucher, 1963). Thus,

TABLE 1. Production of insecticidal substances by varieties of *Bacillus thuringiensis*.

Serotype H[a]	Biotype	Varietal epithets	α-exotoxin	β-exotoxin	γ-exotoxin	δ-endotoxin	Labile exotoxin	Water soluble toxin	Mouse factor exotoxin
1	I	*thuringiensis*	+	+		+	+		+
2	II	*finitimus*	+	-		+			
3a	III 1	*alesti*	+	-		+		+	+
3a,3b	III 2	*kurstaki*	+	-		+			
4a,4b	IV 1	*sotto*	+	-		+			
4a,4b	IV 1'	*dendrolimus*	+	-		+			
4a,4c	IV 2	*kenyae*	+	-		+			
5a,5b	V 1	*galleriae*	-	+		+			+
5a,5c	V 2	*canadensis*	-	+		+			
6	VI	*subtoxicus*	-	-	+	+			
6	VI'	*entomocidus*	-	-	+	+			
7	VII	*aizawai*	+	+		+			
8a,8b	VIII 1	*morrisoni*	-	+		+			
8a,8c	VIII	*ostriniae*	+	-		+			
9	IX	*tolworthi*	+	+		+			
10	X	*darmstadiensis*	-	+		+			
11a,11b	XI 1	*toumanoffi*	+	+		+			
11a,11c	XI 2	*kyushuensis*	+	-		+			
12	XII	*thompsoni*	+	-		+			
13	XIII	*pakistani*	+	-		+			
14	XIV	*israelensis*	+	-		+			
15	XV	*dakota*	+			+			
16	XVI	*indiana*	+			+			
-	-	*wuhanensis*[b]	-	-		+			

[a]Based on classification according to Dr. H. deBarjac, Institut Pasteur, Paris, France.

[b]No flagellar antigen

+ = positive; - = negative; blank = not determined

little is actually known about the epizootiology of most pathogens in natural insect populations of Mediterranean flour moth, *Anagasta kuehniella*, caused by *B. thuringiensis*. In Russia, Talalageva (1967) reported an epizootic of *B. thuringiensis* in field populations of the lepidopteran forest pest, *Selenephera lunigera*. Talalaev (1958, 1959) and Gukasyan (1961) have reported epizootics produced by *B. thuringiensis* in Siberian silkworm, *Dendrolimus sibericus*, populations. Recently, Burges and Hurst (1977) noted a natural epizootic in a population of *Cadra cautella* caused by *B. thuringiensis*.

Epizootics of *B. thuringiensis* is most often associated with man-manipulated environments such as cocooneries (silk-production), and granaries (grain storage). Effectiveness may persist for relatively long periods in dust, grass litter, or insect remains if there is protection from moisture, sunlight, and extreme heat, but *B. thuringiensis* is not a highly infectious natural pathogen. It is, however, relatively cheap to produce *in vitro* and will function as an effective microbial agent if heavily disseminated artificially into an appropriate insect field habitat.

Since *B. thuringiensis* is highly pathogenic to some insect larvae (mainly Lepidoptera) and occurs commonly in nature, one could perhaps expect frequent natural epizootics caused by this bacterium. In fact, persistence and natural spread are relatively poor, even when the material is applied artificially. Therefore, products containing the crystal toxin and spores of *B. thuringiensis* are normally directed toward short-term suppression.

As mentioned above, infection is usually limited to septicemia of a few species such as *Dendrolimus sibericus* (Talalaev, 1958, 1959; Gukasyan, 1961), *Anagasta kuehniella* (Kurstak, 1964), *Selenephera lunigera* (Talalageva, 1967), and *Cadra cautella* (Burges and Hurst, 1977). Thus, it appears to be a natural enzootic regulatory factor. However most varieties of *B. thuringiensis* have been isolated from field insects, which indicates that it is a successful pathogen if persistence, even at a low level, is a measure of success. Obviously, enough spores are produced to ensure persistence, for example, in granaries where the flour moth is present and in cocooneries where silkworm are being reared (Angus, 1968a). Also, in Japan mulberry leaves contaminated with silkworm feces containing *B. thuringiensis* appears to allow for some persistence of the pathogen in the field (T. Iizuka, pers. commun.). Another factor that likely restricts the development of epizootics by *B. thuringiensis* is the low capacity to disperse in the host environment and consequently to infect the hosts. Bacteria are ordinarily dispersed by infected hosts or by healthy insects that harbor an infectious agent (feeding on contaminated food or cannibalism), as is the case with *Coccobacillus acridiorum*,

Pseudomonas aeruginosa, Serratia marcescens, and *B. cereus* in grasshoppers (Bucher and Stephens, 1957; Bucher, 1959; d'Herelle, 1914). Thus, the few unusual epizootics caused by *B. thuringiensis* var. *dendrolimus* probably resulted from the mass migration of the larvae of the Siberian silkworm, *Dendrolimus sibericus*, which distributes the pathogen (Talalaev, 1958, 1959; Gukasyan, 1961).

The pathogenicity to the group of insects which is not sensitive to crystals alone or spores alone, and the sequence of the pathological process due to the ingestion of the spore-crystal mixture is rather complex. Intoxication in this case has been studied by Yamvrias (1961) in *Anagasta kuehniella*. The gut pH of the larvae (approximately 8.0) is not alkaline enough to permit efficient dissolution of the toxic crystals, but it is favorable for spore development. This type of insects in the field is more likely to be susceptible to epizootics. In *Laspeyresia pomonella* the disease is progressive and slow, nonparalytic, and both spores and crystals must be present (Roehrich, 1964). The progression of the disease apparently depends on the dose ingested and on physiological variables which allow the development of the pathogen inside the susceptible host. It has been proposed that a synergistic action of the various toxins is necessary to prolong symptoms and infection (MacEwen et al., 1960). Also, it has been shown that the spores of *B. thuringiensis* remain active in soil for at least 3 months following application (Saleh et al., 1970a), but the concentrations in the soil probably are not sufficient to contaminate insect habitats, nor cause epizootics. Also, the stability of *B. thuringiensis* spores in soil, though it is not affected by pH of crop soils (Saleh et al., 1970a), is affected by pH of organic amendments (Saleh et al., 1970b). In addition, Smirnoff (1968) found that foliage of some plant species released substances that were bacteriostatic for *B. thuringiensis*. Lastly, the failure of *B. thuringiensis* to cause epizootics under natural conditions may relate to the fact that dead infected insects contain large numbers of vegetative cells but only a few spores and crystals. Thus, the principal agents of mortality do not become abundant rapidly in an affected insect population.

IV. INSECTICIDAL TOXINS OF *Bacillus thuringiensis*

B. thuringiensis α-exotoxin. This toxin (phospholipase C, lecithinase) has been studied in connection with the pathogenicity of both *B. thuringiensis* and *B. cereus*. The enzyme acts upon many kinds of cells, primarily by affecting the phospholipid of the membranes and causing lytic or necrotizing action. It acts on the lecithin molecule with the formation of a diglyceride and phosphorylcholine.

A comparative study (Ivinskene et al., 1975) of the lecithinase biosynthesis by *B. thuringiensis* on four nutrient media showed that lecithinase appears in the logarithmic phase of growth and that the greatest lecithinase activity in the culture fluid is observed 10 hr after inoculation of the microbe (at the beginning of the stationary phase). The biosynthesis and accumulation of lecithinase occur in the range of pH 6-9. Lecithinase is a thermolabile protein that is stable over a broad range of pH (3-9) and stable in the presence of trypsin and 8*M* urea.

Toumanoff (1953) was the first to point out the similarity between the lecithinase produced by *B. thuringiensis* var. *alesti* and the toxin of *Clostridium perfringens*. Later, Heimpel (1954a,b; 1955a,b) demonstrated the coincidence between the pH optimum (6.6-7.4) for lecithinase and the pH inside the gut of sawfly larvae. Species of sawflies whose gut pH was higher than the optimum for the enzyme were not susceptible to *B. cereus* and histological examination of the insect gut did not reveal enzymic destruction; also, species of *Bacillus* that are incapable of producing lecithinase were not found to be pathogenic to the larch sawfly. In addition, when *B. cereus* was tested against other insect species, it was found that resistant species had an alkaline pH in the midgut. Thus, the alkaline condition in the gut of some insects seems to be a limiting factor in both growth of the organism and lecithinase activity. Toumanoff (1954) observed a toxic reaction of larvae of *Galleria mellonella* after application of precipitated protein from culture filtrates of *B. thuringiensis* var. *alesti*; the filtrates contained phospholipase C and the toxicity was interpreted as a result of activity of that enzyme. Krieg (1971) confirmed that supernatant fluids of cultures of *B. thuringiensis* containing phospholipase C may induce a toxic reaction in such insect larvae as *Plutella maculipennis* after aplication per os. However, the role of this enzyme from *B. thuringiensis* in pathogenicity is not known and needs further confirmation.

B. thuringiensis β-exotoxin. In 1959-60, certain varieties of *B. thuringiensis* were reported to produce an exotoxin (*B. thuringiensis* β-exotoxin, fly factor, fly toxin, thermostable exotoxin, McConnell and Richard's factor) that was deleterious to dipteran larvae and pupae and to some Lepidoptera (McConnell and Richards, 1959; Burgerjon and deBarjac, 1960a,b). This material was characterized as being heat-stable, secreted to the outside of the bacterial cell into the culture medium during the active phase of vegetative growth, water soluble and dialyzable, absorbent at 260 mμ, and nucleotide-like in structure. It has been reported to be produced by serotypes 1, 5a/5b, 5a/5c, 7, 8a/8b, 9, 10, and 11a/11b of *B. thuringiensis* (Cantwell et al., 1964;

Burgerjon and deBarjac, 1967a; Heimpel, 1967; deBarjac et al., 1966; Krieg et al., 1968).

Several attempts have been made to purify and identify the *B. thuringiensis* β-exotoxin (deBarjac and Dedoner, 1965, 1968; Bond et al., 1969; Sebesta et al., 1969a,b; Farkas et al., 1969). Normally, *B. thuringiensis* var. *thuringiensis* produces approximately 50 mg of *B. thuringiensis* β-exotoxin/liter of supernatant liquid. For purification, the vegetative cells, spores, and crystals are centrifuged into a pellet, and the supernatant fluid is sterilized for 15 min at 120°C. Then the *B. thuringiensis* β-exotoxin can be adsorbed onto acid-washed charcoal, eluted from the charcoal with 50% ethanol, and concentrated. The concentrated material can now be chromatographed on paper by using ethanol-1*M* ammonium acetate (pH 3.8) or fractionated by using Dowex-1 formate and eluting the material from the column with 0-1.5*M* ammonium formate buffer, desalting on Sephadex, and bioassaying for toxicity. Remaining impurities can be removed by further chromatography with resins such as DEAE cellulose. Sebesta et al. (1969) and Farkas et al. (1969) identified the *B. thuringiensis* β-exotoxin structurally from a strain of *B. thuringiensis* (designated by them as var. *gelechiae*). The molecular weight is calculated from the proposed structure at 699.35 and contained, in addition to equimolecular amounts of adenine, ribose, and phosphorus, a sugar moiety possessing two carboxylic groups which can form five-membered lactones with adjacent hydroxylic functions. The phosphoric acid residue is bound to the hydroxylic function of one from the two *cis* diol systems present. Thus, the structure is a nucleotide containing adenine linked to ribose which in turn is linked to glucose (4",4). The glucose is linked to the dicarboxylic moiety (allomucic acid) by a 1,2" bond with substitution of a phosphate residue at position 4" of the allomucic acid.

The metabolic pathway for production of the *B. thuringiensis* β-exotoxin is not actually known. However, when Shieh et al. (1968) studied nucleotide metabolism in *B. thuringiensis* var. *thuringiensis* by using cell-free extracts and adding certain purine derivatives, the purine additives decreased the accumulation of *B. thuringiensis* β-exotoxin, which indicated that normal feedback mechanisms were operating. Also, the catabolic pathways involved in the degradation of nucleotides by nucleotidase were operating. However, the *B. thuringiensis* β-exotoxin was resistant to nucleotidase degradation and, in fact, it was inhibitory to this enzyme. Thus, when resting cells of the organisms were incubated with adenosine monophosphate (AMP) or inosine monophosphate (IMP), the release of inorganic phosphorus from AMP was almost completely inhibited, and its release from IMP was partially inhibited. Therefore, there is probably a decrease in rate of the catabolic formation of purine bases from their respective nucleotides when the *B. thuringiensis* β-exotoxin is used as a nucleotidase inhibitor. This lower concentration of adenine and guanine means that purine monophosphate synthesis can occur continually without feedback and that IMP, for example, can be directed through the AMP branch of synthesis. Then, since AMP is presumably removed in the synthesis of the *B. thuringiensis* β-exotoxin, the equilibrium is shifted toward AMP formation, and a *B. thuringiensis* β-exotoxin-accumulating cell appears to mimic a catabolically blocked mutant (Bond et al., 1971).

The *B. thuringiensis* β-exotoxin is characterized by its ability to prevent pupae of treated larvae of houseflies, *Musca domestica,* from developing into normal, complete adults, and it appears to act whenever cell mitosis occurs at molting or during metamorphosis; thus the effect is often delayed. (It can cause death during molting, completely prevent pupation, or cause chronic poisoning in adults). For example, when housefly larvae receive sublethal doses, vestigial wings and narrow, pointed abdomens develop in the adult (Cantwell et al., 1964).

Burgerjon et al. (1969) found that the effect of *B. thuringiensis* β-exotoxin in the Colorado potato beetle, *Leptinotarsa decemlineata,* involved the buccal parts, eyes, and antennae and that these organs atrophied or morphogenetic anomalies such as clubs or claws occurred in the antennae. (The claws were similar to those found normally on the tarsae of the legs.) Also, protuberances were found on the eyes and the palps and transformation of the paired part of the labial palps into an unpaired part was noted. Thus, in susceptible insects, the least of the teratological effects appears to be atrophy of the mouthparts, since the first part attacked was the proboscis of pupal and adult Lepidoptera and the labrum of the housefly. In *Pieris brassicae,* atrophy of the proboscis prevented feeding (Burgerjon and Biache, 1967a). Table 2 lists the susceptibility of some organisms to the β-exotoxin produced by varieties of *B. thuringiensis.*

B. thuringiensis γ-exotoxin. As mentioned earlier, the *B. thuringiensis* γ-exotoxin is an unidentified enzyme(s) responsible for clearing egg yolk agar (Heimpel, 1967). No proof of toxicity has been obtained and the precise nature of the *B. thuringiensis* γ-exotoxin has yet to be determined.

B. thuringiensis δ-endotoxin. This toxin is contained in a protein parasporal crystal that is thermolabile and soluble in alkaline solutions and is formed in the vegetative cells at the same time the spore is produced. Except in *B. thuringiensis* isolate BA-068 (Reeves and Garcia, 1971a,b), *B. thuringiensis* var. *israelensis* (Lüthy, 1980), *B. thuringiensis* var. *darmstadiensis* (Krieg et al., 1968; Krieg, 1969), which are biocrystalliferous, and a strain of *B. thuringiensis* var. *kurstaki* that produces 2-5 crystals/cell (S. Amonkar, pers. commun.) only 1 crystal/vegetative cell is usually produced. However, the temporal relationship of the formation of this crystal during sporulation sometimes varies among strains as does the site of synthesis within the cell. For example, the parasporal crystal is often first observed during stage III of sporulation (engulfment) (Bechtel and Bulla, 1976) and at about the same time, an ovoid inclusion appears (although more than one crystal can develop within a cell, only one ovoid inclusion per sporangium is ever observed). This ovoid inclusion is never observed to exhibit crystal lattice fringes as does the parasporal crystal, and it is not known whether the ovoid inclusion possesses insecticidal properties.

Interestingly, considerable evidence has accumulated (Somerville and Pockett, 1976; Lecadet and Dedonder, 1971; Ribier, 1971; Lecadet et al., 1972; Ribier and Lecadet, 1973; Herbert and Gould, 1973) indicating a relationship between the protein of the crystal of *B. thuringiensis* and the spore material that is solubilized under conditions similar to those necessary to dissolve crystals. Also, toxin can be extracted from the spores by using a reducing buffer. Somerville and Pockett (1975) revealed that spore and crystal toxins share common antigens and concluded that the spore toxin is located in the exosporium and spore coat. Scherrer et al. (1974) presented results confirming the location of toxicity in the exosporium.

Electron microscopy of sections and of ultrasonic extracts confirmed that the crystalline inclusions appear during a defined phase of the sporulation process and have not only a regular form, but also a fine striated surface. At the early stages of formation they show rudiments of the final shape and surface. Labaw's (1964) work with *B. thuringiensis* var. *thuringiensis* in which he used electron micrographs of shadowed carbon replicas of parasporal crystals revealed that the protein crystal is a tetramolecular face-centered cubic unit cell 123 Å on an edge. The protein molecules appeared to be spherical and had a diameter of 85 Å. The bipyramidal shape of these crystals, according to Labaw, results from their being bounded by 8 similar faces on which the rows of molecules are separated by 3 diameters. Holmes and Monro (1965) then expanded on this work and estimated by deduction a molecular weight of about 230,000 for the unit. Grigorova et al. (1967) described two types of crystals (bipyramidal and biprismatic) and suggested on the basis of measurements of a number of crystals that the molecules of both types have the form of a rotary

TABLE 2. Susceptability[a] of certain organisms to the β-exotoxin produced by varieties of *B. thuringiensis*.[b]

Phylum	Organisms	Injection		Feeding
Annelida	*Tubifex* sp.			0
Arthropoda	*Aedes aegypti* (L.)	0		+
	Apis mellifera L.	++		+
	Blatta orientalis L.	++		
	Bombyx mori (L.)	++		+
	Diprion pini (L.)	++		+
	Drosophila melanogaster Meigen			+
	Estigmene acrea (Drury)	0		+
	Euxoa segetum (Denis & Schiffermuller	0 to +		+
	Galleria mellonella (L.)	++		+
	Lepinotarsa decemlineata (Say)	++		+
	Locusta migratoria (L.)	++		+
	Lymantria dispar (L.)	+		+
	Malacosoma neustria (L.)	++		+
	Peridroma saucia (Hübner)	0 to +		+
	Phaenicia sericata (Meigen)	++		+
	Pieris brassicae (L.)	+		+
	P. oleracea Harris	++		+
	Musca autumnalis De Geer	++		+
	M. domestica L.	++		+
	Ostrinia nubilalis (Hübner)	++		+
	Periplaneta americana (L.)	0		+
	Plodia interpunctella (Hübner)	0		+
	Plutella maculipennis (Curtis)			+
	Pristiphora pallipes Lepeletier	++		+
	Reticulitermes flavipes (Kollar)			+
	Reticulitermes hesperus Banks			+
	R. virginicus (Banks)			+
	Sarcophaga bullata Parker	++		
	Tetranychus urticae Koch			+
	Zootermopsis angusticollis (Hagen)			+
Chordata	*Mus musculus* (L.)	+		
			Ova	
Nemathelminthes	*Bunostomum trigonocephalum* (Rudolphi)		+	
	Chabertia ovina (Fabricius)		+	
	Cooperia (Ransom)		+	
	Haemonchus contortus (Rudolphi)		+	
	Oesophagotomum venulosum (Rudolphi)		+	
	Ostertagia (Ransom)		+	

[a] ++ = high toxicity; + = low toxicity; + to 0 = variable; 0 = no toxicity.

[b] Compiled from Heimpel (1967) and Bond et al. (1971).

ellipsoid. Finally, descriptions of rhomboid, cuboidal, truncated, or wedge-shaped crystals have been published (Hannay and Fitz-James, 1955; Steinhaus, 1954; Bucher et al., 1966; Fujiyoshi, 1973). Scherrer et al. (1973) reported that crystal shape is influenced by components of medium since levels of glucose above 0.8% caused crystals of *B. thuringiensis* var. *thuringiensis* to be amorphous rather than bipyramidal.

The parasporal crystal of *B. thuringiensis* is made up of a number of distinct polypeptide components but the exact number and their properties are not well defined because buffers at alkaline pH or thiol reagents that can cause peptide and amide bond cleavage are used to solubilize the crystal. Consequently, there is uncertainty concerning the biochemical and biophysical properties. Although molecular weights that range from 500 to over 200,000 have been calculated for major components derived from the parasporal crystal, these differences have been attributed to differences in the strains analyzed and in the methods used for solubilization (Somerville, 1973, 1977). The results obtained by various workers indicate that of most of the methods of solubilizing parasporal crystals produce a toxic, high molecular weight complex and a toxic low molecular moiety, but depending upon the conditions of crystal dissociation and characterization, the two complexes may break down to relatively smaller units. In addition, the relationship between the two units is not yet clear. Possibly, the small molecular weight complex is generated from the larger one. [I have used the term "complex" because there is some indication from the detailed studies of Lecadet and Dedonder (1967) and Lacadet and Martouret (1962, 1967) that a mixture or complex may be more toxic than individual proteins and peptides. For example, toxicity decreased when they purified the various proteinaceous components.]

Cooksey (1968) hydrolyzed crystals with gut juice enzymes from *Pieris brassicae*, and with chymotrypsin, trypsin, and subtilisin. He isolated a toxic and antigenic component with an apparent molecular weight of about 40,000; other components with small molecular weights (~10,000) were not antigenic. Pendleton (1968) also treated crystals with gut juice of *P. brassicae* and examined the resulting products by gel filtration, antigenic analysis, electrophoresis, and insect bioassays. As a result, he separated the gut enzyme digests into two fractions, one of apparent molecular weight of ~200,000 and the second of 5,000-10,000. Fraction one was toxic by ingestion alone; fraction two was toxic upon injection and ingestion. Fast and Angus (1970) reported the presence of a toxic peptide of molecular weight 500-1000 obtained by dissolving crystals in 0.04 *M* NaOH and 0.2 *M* β-mercaptoethanol (ME), filtering through a membrane, and incubating with trypsin. Sayles et al. (1970) also dissociated the crystal into several polypeptides with molecular weights of approximately 1,000 by using 8 *M* urea (pH 6.0), 0.5% dithiothreitol, and gel filtration chromatography. Whether the components were insecticidal was not determined. Later work by Aronson and Tillinghast (1976) corroborated the presence of small molecular weight components in material dissolved under denaturing conditions. Glatron et al. (1972) made a systematic study of the release of soluble protein from the crystalline inclusion by using ME at pH 7.5 and 9.5 with and without 6 *M* guanidinium chloride and NaOH. Solubilization in NaOH caused the release of four amino-terminal residues; solubilization in ME and guanidinium chloride produced only one amino-terminal amino acid (phenylalanine), an indication of a single component with a molecular weight calculated to be 80,000.

Faust et al. (1974a) examined crystal degradation products from *B. thuringiensis* var. *kurstaki* obtained by treatment with alkali or gut juice from larvae of *Bombyx mori*, and various plant and mammalian enzymes. Highly toxic material was associated with a protein having a molecular weight of 230,000. Other toxic components produced ranged in molecular size from 235,000 to 1000. Herbert et al. (1971) separated two polypeptide chains having apparent molecular weights of 55,000 and 120,000 from crystals solubilized in "Ellis" buffer. These investigators postulated that the 120,000 dalton protein is a dimer of the smaller one. Curiously, the smaller polypeptide was toxic; the larger one was not. Prasad and Shethna (1974) used 1 *N* NaOH for crystal solubilization, and followed it by anion exchange chromatography and gel filtration; they obtained two components toxic to *B. mori*.

From careful analysis of the above results one can reason that the crystal is composed of a single subunit that is converted to smaller components by the usual conditions of dissolution Compelling evidence for this hypothesis was presented by Chestukhina et al. (1977) and by Bulla et al. (1977). The former investigators demonstrated the presence of a major component (M.W. $\cong$ 130,000) upon selective extraction with reducing and denaturing agents. Their results agree well with those of Herbert et al. (1971). Bulla et al. (1977) then demonstrated that the parasporal crystal of *B. thuringiensis* is composed of a single glycoprotein subunit with a molecular weight of about 90,000-130,000; carbohydrate constitutes about 5% of the crystal, and the balance is protein.

The amino acid and carbohydrate composition of several varieties of *B. thuringiensis* δ-endotoxin, as determined by investigators, are listed in Table 3. Glutamic acid and aspartic acid residues are the most abundant; glucose (3.8%) and mannose (1.8%) account for all of the carbohydrate present; and on the basis of molecular weights, there are 20 glucose and 10 mannose

TABLE 3. Chemical composition of the δ-endotoxin from several varieties of *Bacillus thuringiensis*.

Amino acid or sugar	Percent							
	kurstaki[a/]	*thuringiensis*[b/]	*alesti*[c/]	*entomocidus*[c/]	*kenya*[c/]	*tolworthi*[d/]	*galleriae*[d/]	*sotto*[e/]
Aspartic acid	12.4	11.9	13.0	14.0	12.4	13.2	14.4	9.6
Threonine	5.5	5.2	7.0	6.8	6.3	5.3	6.6	4.5
Serine	5.9	5.4	6.5	6.0	5.3	5.6	8.3	4.8
Glutamic acid	13.1	12.5	17.0	16.0	13.7	10.7	11.7	11.8
Proline	2.8	4.0	4.3	4.4	3.4	3.9	5.6	7.5
Glycine	3.6	3.0	4.8	4.8	3.6	7.6	6.0	3.2
Alanine	3.3	3.1	5.0	4.8	4.0	5.5	6.4	2.8
Half-cystine	1.4	—	—	—	—	1.1	1.3	—
Cystine	—	0.3	1.9	1.4	2.0	—	—	1.2
Valine	5.8	5.7	7.4	7.8	5.8	7.4	4.4	5.3
Methionine	0.9	1.6	0.2	0.7	0.4	0.7	1.2	1.3
Isoleucine	5.4	6.3	6.4	6.6	5.6	6.2	5.6	—
Leucine	7.7	8.8	9.5	10.5	8.3	8.2	9.0	11.2
Tyrosine	5.7	5.9	6.7	6.5	5.8	3.7	4.3	6.8
Phenylalanine	5.0	5.2	6.8	6.7	5.9	4.5	4.1	8.6
Lysine	2.8	4.0	5.4	4.4	4.6	4.2	2.8	3.6
Histidine	1.9	2.5	2.8	2.8	2.8	2.5	2.2	2.7
Arginine	10.3	8.5	10.4	9.0	8.6	8.3	6.2	9.6
Tryptophan	1.7	5.2	2.3	2.3	—	—	—	2.6
Glucose	3.8	—	—	—	—	—	—	—
Mannose	1.8	—	—	—	—	—	—	—

a/ Bulla et al., 1977
b/ Holmes and Monro, 1965
c/ Spencer, 1968
d/ Chestukhina et al., 1978
e/ Angus, 1956c.

residues per subunit. Bateson and Stainsby (1970) reported the presence of these hexoses as well as the pentoses arabinose and xylose and verified that there were not detectable amino sugars as was reported previously (Akune et al., 1971). Neither were there any lipids, nucleic acids, or sialic acid derivatives. The structure of the individual carbohydrate side chains or their specific site(s) of attachment to the protein are now known. Finally, elemental analysis has revealed the presence of Ca (0.75%), Fe (0.14%), Mg (0.42%), and Si (0.32%) (Faust et al., 1973), but the role, if any, of these elements in the overall structure of the crystal also is not known.

Treatment of crystals with excess alkali reduces the yield of soluble protoxin, as measured by insect toxicity and gel filtration chromatography (Bulla et al., 1977). For example, no biologically active protein is recovered from crystals treated with 1 *N* NaOH for 20 hr. Time of incubation also has an effect. Dissolution is complete in 13.5 m*M* NaOH after 3 hr; however; about 40% of the subunit remains in an aggregated state as revealed by anion-exchange chromatography. Only after 4-5 hr is the aggregated protein nearly completed dissociated. After more than 24 hr at pH 12, the subunit yield decreases, and after 168 hr, only 30% of the subunit remains. Since biological activity, i.e., insect toxicity, of alkali-treated crystals is directly proportional to the protoxin yield, as determined by anion-exchange chromatography, solubility, and toxicity of the crystal are obviously both pH- and time-dependent. Nishiitsutsuji-Uwo et al. (1977) made similar observations regarding pH: maximum solubility was obtianed after about 5 hr when the crystal was titrated with 400 equivalents of base. The subunit was stable for several hours thereafter, but it eventually degraded to smaller fragments and showed a concomitant loss in insecticidal activity. However, reaggregation also occurred, especially when the pH was lowered to neutrality.

Several investigators have recently reported results that may lead to a clearer understanding of the genetic mechanism controlling crystal production. There is some evidence that plasmid(s) may be involved in the synthesis of the parasporal crystals responsible for the pathogenicity of *B. thuringiensis* to insects (Debabov et al., 1977; Ermakova et al., 1978; Galushka and Azizbekyan, 1977; Stahly et al., 1978b). For example, Stahly et al. (1978b) examined crystalliferous and acrystalliferous strains of *B. thuringiensis* var. *kurstaki* and found that crystalliferous strains contained at least six extrachromosomal DNA molecules that ranged from 1.32 to 47.13 x 10^6 daltons. All nontoxic acrystalliferous mutants lacked the complete array present in the wild-type toxic strains, an indication of relationship between the presence of plasmid(s) and toxicity. Furthermore, the very high frequency of the acrystalliferous mutants suggested involvement of an unstable genetic element such as a plasmid.

Miteva (1978) examined 14 strains of *B. thuringiensis* and 6 strains of closely related *B. cereus* for the presence of plasmid DNA. Results of the electrophoretic analysis demonstrated that all the *B. thuringiensis* strains possessed covalently closed circular double-stranded DNA. Plasmid DNA was present in 3 of the 6 strains of *B. cereus*. However, the plasmids isolated from some asporogenous and acrystalline mutants of *B. thuringiensis* appeared no different from those of the parent strains. The fact that the asporogenous and acrystalline mutants retained both their plasmids and their ability to form crystals does not exclude the possibility that plasmids of *B. thuringiensis* take part in the genetic determination of crystal formation; the loss of protein crystals in Miteva's study might have been the result of a mutation.

Ermakova et al. (1978) demonstrated that when *B. thuringiensis* var. *galleriae* was grown in selective media that inhibited crystal formation, no extrachromosomal DNA could be isolated from the cells. Thus these results suggested a correlation between the presence of plasmid DNA, the formation of crystals, and the level of intracellular protease activity. Also, the results of medium-shift experiments in their work support their hypothesis that plasmid DNA may have a chromosomal origin, that is, it may result from specific excision and amplification of certain chromosomal DNA segments. This explanation is also consistant with previously reported experiments with *B. megaterium* plasmids (Carlton 1976). Galuska and Azizbekyan (1977) found that no extrachromosomal DNA elements could be detected after certain *B. thuringiensis* strains were cured with ethidium bromide or cultured under extreme conditions (increased temperature, alkaline pH of medium) and selected for acrystalline strains. Debabov et al. (1977) obtained similar results.

Other researchers have reported plasmid DNAs in *B. thuringiensis*. Stahley et al. (1978a) examined extrachromosomal DNA in *B. thuringiensis* var. *alesti* by electron microscopy and identified 12 size classes of molecules ranging from 2.60 to 44.58×10^6 daltons. Debabov et al. (1977) examined *B. thuringiensis* var. *galleriae* and found three kins of circular DNA molecules with sizes of 6, 10, and 12×10^6 daltons. Ermakova et al. (1978) examined *B. thuringiensis* var. *galleriae* and also found three plasmids with molecular weights of 5.9, 10, and 10.9×10^6 daltons. Galushka and Azizbekyan (1977) examined *B. thuringiensis* var. *thompsoni*, var. *tolworthi*, and var. *finitimus* and found plasmid DNA in these varieties, but did not report molecular weight sizes. It appears, therefore, that the varietal types of *B. thuringiensis* exhibit different plasmid profiles.

Faust et al. (1979) recently examined four entomopathogenic bacteria for extrachromosomal DNA molecules. The basic characteristics of the larger isolated elements, especially the giant DNA elements, were not unlike those of representative plasmids isolated from members of other genera of bacteria (Clowes, 1972). *B. thuringiensis* var. *kurstaki* contained 12 elements that banded on agarose gels; these ranged from 0.74 to >50 x 10^6 daltons, and three were giant extrachromosomal DNA elements. *B. thuringiensis* var. *sotto* contained one giant extrachromosomal DNA element with a molecular size of about 23.5 x 10^6 daltons and two lesser elements of 0.8 and 0.62 x 10^6 daltons. *B. thuringiensis* var. *finitimus* contained two giant DNA elements corresponding to >50 x 10^6 daltons and two lesser elements of relatively small size (0.98 and 0.79 x 10^6 daltons). *B. popilliae* contained no giant extrachromosomal DNA elements but did contain two smaller DNA elements corresponding to 4.45 and 0.58 x 10^6 daltons.

Also, in a recent study Iizuka et al. (1981a) examined all 17 serotypes (24 strains) of *B. thuringiensis* for the presence of extrachromosomal DNA by agarose gel electrophoresis. The number of plasmid bands based on the electrophoresis profiles ranged from one for *B. thuringiensis* var. *sotto*, and var. *thompsoni* to 16 for *B. thuringiensis* var. *kurstaki*. The reported plasmid profiles consisted of both covalently closed circular (ccc) and open circular (oc) forms with molecular weights of the ccc forms ranging from less than 1 megadalton to greater than 200 megadaltons, depending on the strain. In serologically identical varieties (*thuringiensis* Berliner and *thuringiensis* BA-068, serotype 1; *sotto* and *dendrolimus*, serotype 4a, 4b; *subtoxicus* and *entomocidus*, serotype 6), only serotype 6 strains showed similar extrachromosomal DNA profiles and while a number of strains of *B. thuringiensis* have some extrachromosomal DNA elements in common, distinct differences can be observed in the DNA profiles between serotypes and even among strains of the same serotype.

A concomitant study by the same research team (Iizuka et al., 1981b) compared the plasmid profile of single and multiple crystalliferous strains of *B. thuringiensis* var. *kurstaki*. Plasmid DNA from the multicrystalliferous strain (2-5 crystals/cell) was compared to *B. thuringiensis* var. *kurstaki* (HD-1 and HD-73) in an attempt to correlate the presence of plasmids with the production of parasporal crystals and to ascertain whether or not there is a correlation between plasmid profiles and strains of the same serotype. An 18.62 megadalton plasmid was seen in the multicrystalliferous strain that was not detected in HD-1 or HD-73 strains of *B. thuringiensis*. However, other differences in the plasmid profile made it futile to associate plasmid DNA with crystal production. Further genetic and biochemical studies are obviously necessary to determine the biological function of extra-chromosomal DNA elements in *B. thuringiensis* and to definitely determine which are indeed autonomous replicons that may be useful in genetic manipulation to improve effectiveness of these organisms.

The *B. thuringiensis* δ-endotoxin is mainly toxic for lepidopteran larvae though other types of insects including larvae of Japanese beetles, *Popillia japonica* (Sharpe, 1976), blackflies (Lacye and Mulla, 1977), and mosquitoes (deBarjac, 1978a,b,c; Fujiyoshi, 1973; Reeves and Garcia, 1971a,b; Hall et al., 1977) are susceptible.

Once a susceptible insect ingests parasporal crystals, it stops feeding, becomes dehydrated, and dies, but death usually is preceded by sloughing of the epithelial cells that border the midgut. Since the midgut epithelium of insect larvae is critical to ionic regulation of the hemolymph, particularly potassium ions, it is little wonder that inhibition of ionic control facilitates toxic effects on the insect.

Insects of the order Lepidoptera were placed by Heimpel and Angus (1963) into four groups according to their response to *B. thuringiensis*. Those that exhibit general paralysis after ingestion of the δ-endotoxin and show a blood pH change are type I insects, e.g., *Bombyx mori*. A type II insect is representative of most Lepidoptera: these insects suffer gut paralysis shortly after feeding on the δ-endotoxin, but there is no gut leakage and, consequently, no change in blood pH or general paralysis; the insects starve and die from a bacteriaremia. Type III insects succumb to a normal whole preparation of *B. thuringiensis:* both spores and δ-endotoxin must be present to cause death, but large doses of β-exotoxin (fly toxin) alone can kill. This group is represented by *Anagasta kuehniella* and *Lymantra (=Porthetria) dispar* (Burgerjon and deBarjac, 1960b). Type IV insects were described by Martouret (1961); they are those Lepidoptera that are not susceptible to the δ-endotoxin. Examples would be noctuids such as *Mamestra brassicae*.

The susceptibility of a specific insect to a specific strain of *B. thuringiensis* has been reported (van der Laan and Wassink, 1964; Vankova, 1964); the activity of several strains against a specific insect (Dulmage et al., 1981) or the activity of several strains against several insects (Burgerjon and deBarjac,

1964; Burgerion and Biache, 1967; Grigorova, 1964). In most instances tests for insecticidal activity are confined to testing known β-exotoxin- or δ-endotoxin-producing strains against insects that are susceptible to β-exotoxin or δ-endotoxin.

However, results of tests of whole cultures of *B. thuringiensis* varieties against larvae of the silkworm, *B. mori* indicated that a gradation of toxicity exists (Angus and Norris, 1968). A scale of toxicity can therefore be devised as follow: very high (less than 0.25 μg/g); high (less than 1.0 μg/g); moderate (1-3 μg/g); low (3-10 μg/g); and very low (more than 10 μg/g). The varieties *finitimus* and *subtoxicus* were not toxic at the levels tested (up to 100 μg/g). *B. thuringiensis* var. *thuringiensis* had very low toxicity as did *B. thuringiensis* var. *dendrolimus*. *B. thuringiensis* var. *tolworthi* had moderate toxicity. *B. thuringiensis* var. *morrisoni* was highly toxic, and *B. thuringiensis* var. *alesti, entomocidus,* and *sotto* were very highly toxic. On the other hand, the findings of Angus and Norris (1968) and Rogoff et al. (1969) indicate that within a given serotype there is a considerable variation in toxin-forming ability under a given condition of growth. In fact, careful examination of the Angus and Norris (1968) studies also showed a gradation of toxicities both within and between serotypes. For example, they reported a toxicity (ED_{50} dose, paralysis in 12 hr) for crystals of the 562-5A strain of a serotype 1 culture as 0.2 μg/g larval weight, but for the serotype 1 Mattes strain, a dose of 65 μg/g was reported. Not only do these doses lie at extreme ends of the scale for all serotypes tested, but they represent a 300 fold difference in toxicity within the same serotype grouping. Rogoff et al. (1969) reported similar findings with other test insects. Probably the most extensive testing of *B. thuringiensis* serotypes and strains within serotypes has been conducted by Dulmage and his colleagues in the International Cooperative Program on the Spectrum of Activity of *Bacillus thuringiensis*. They have found tremendous variations in spectra of activity and toxicity between various strains and have reported their findings in a recent detailed review to which the reader is referred (Sulmage et al., 1981).

The reactions of susceptible insects to the strains and to the *B. thuringiensis* δ-endotoxin of *B. thuringiensis* are likewise rather variable. Therefore, the choice of experimental conditions such as the criteria of interpretation, the doses used, and the age of the test insects can have profound effects on the results. Table 4 lists some of the lepidopterous insects that are susceptible to the *B. thuringiensis* δ-endotoxin produced by *B. thuringiensis* strains.

TABLE 4. A List of Some Lepidopterous Insects That are Susceptible to the δ-Endotoxin Produced by Several Varieties of *Bacillus thuringiensis*.[a]

SCIENTIFIC NAME	COMMON NAME
Achroia grisella	Lesser wax moth
Agrostis ipsilon	Black cutworm
Alabama argillacea	Cotton leafworm
Alsophila pometaria	Fall cankerworm
Anagasta kuehniella	Mediterranean flour moth
Anarsia lineatella	Peach twig borer
Antheraea pernyi	Sinjyu-silkworm
Archips argyrosphilus	Fruittree leafroller
Argyrotaenia velutinana	Redbanded leafroller
Bombyx mori	Silkworm
Choristoneura fumigerana	Spruce budworm
Choristoneura rosaceana	Obliquebanded leafroller
Colia eurytheme	Alfalfa caterpillar
Crambus sperryellus	Lawn moth
Datana integerrima	Walnut caterpillar
Desmia funeralis	Grape leaffolder
Diatera saccharalis	Sugarcane borer
Erannis tiliaria	Linden looper
Estigmene acrea	Saltmarsh caterpillar
Galleria mellonella	Greater wax moth
Heliothis virescens	Tobacco budworm
Heliothis zea	Bollworm
Hemileuca oliviae	Range caterpillar
Hyphantria cunea	Fall webworm
Laspeyresia pomomiella	Codling moth
Lymantria dispar	Gypsy moth
Malacosoma americanum	Eastern tent caterpillar
Malacosoma californicum	Western tent caterpillar
Malacosoma disstria	Forest tent caterpillar
Malacosoma californicum fragile	Great Basin tent caterpillar
Manduca quinquemaculata	Tomato hornworm
Manduca sexta	Tobacco hornworm
Nygmia phaerorrhoea	Browntail moth
Operophtera brumata	Winter moth
Orgyra vetusta	Western tussock moth
Ostrinia nubilatis	European corn borer
Paleacrita vernata	Spring cankerworm
Papilio cresphontes	Orangedog
Pectinophora gossypiella	Pink bollworm

TABLE 4. (Continued)

SCIENTIFIC NAME	COMMON NAME
Phryganidia californica	California oakworm
Pieris brassicae	Large white butterfly
Pieris rapae	Inported Cabbageworm
Plathypena scabra	Green cloverworm
Platyptilia carduidactyla	Artichoke plum moth
Plodia interpunctella	Indianmeal moth
Plutella xylostella	Diamondback moth
Proxenus mindara	Roughskinned cutworm
Pseudaletia unipuncta	Armyworm
Spodoptera exigua	Beet armyworm
Spodoptera frugiperda	Fall armyworm
Spodoptera praefica	West yellowstriped armyworm
Stilpnotia salicis	Satin moth
Thymelicus lineola	Introduced European skipper
Thyridopteryx epyemeraeformis	Bagworm
Tineola bisselliella	Webbing clothes moth
Trichoplusia ni	Cabbage looper
Udea ribigalis	Celery leaftier

[a]Compiled from Heimpel and Angus (1963) and Burges and Hussey (1971).

Other toxins. Smirnoff and Berlinguet (1966) reported a new undescribed toxin produced by commercial preparations of *B. thuringiensis* var. *thuringiensis* that was toxic to larvae of Tenthredinidae. (The toxicity to 19 species of sawflies varied with the species and the larval instar tested). Because the toxic substance is very labile (sensitive to air, sunlight, oxygen, 40°C for 144-264 hr, and temperatures above 60°C for 10-15 min), it was called the "labile exotoxin" to differentiate it from the other toxins isolated from *B. thuringiensis*. Preliminary biochemical analyses of the substance revealed that it is composed of 17 amino acids and characterized by the presence of one or more peptides of low molecular weight. It is unlike the *B. thuringiensis* δ-endotoxin in that it contains 5.7% aspartic acid rather than the 16-17% glutamic and 12-15% aspartic acids that the parasporal crystals contain. The other amino acids also are much lower in content. Its mode of action has not been determined, and it is uncertain whether it is actually produced by *B. thuringiensis* or is an artifact of the fermentation process.

Fast (1971) isolated a water-soluble toxin from a commercial preparation of *B. thuringiensis* var. *alesti* that within 3 hr of dosing (*Bombyx mori* larvae) evoked symptoms very similar to those of the parasporal crystal toxin: cessation of feeding, absence of response to external stimuli, and flaccid paralysis. Its molecular weight was found to be greater than 30,000. However, serological studies suggest that it may not be related to the *B. thuringiensis* δ-endotoxin. The toxin may be of use in biological control, but much more data are required.

Krieg (1971) reported a thermo-sensitive "mouse-factor" exotoxin of proteinaceous chacter that is produced during growth phase by strains of *B. thuringiensis* (and *B. cereus*) and is highly effective against mice and several lepidopteran insects. (Sub-lethal doses of the "mouse-factor" caused reduction of growth and prolongation of development). This toxic factor has been related to Heimpel's *B. thuringiensis* δ-exotoxin, though it is not identical with lecithinase C, i.e., the reaction was not limited to lecithinase positive strains. Its importance in pathogenicity is not known.

Smirnoff and Valero (1977) carried out studies to determine if one or more of the varieties of *B. thuringiensis* had chitinolytic enzymatic properties and to obtain more information on the pathogenic activity of *B. thuringiensis* on lepidopterans. Their interest was stimulated by a previous investigation (Smirnoff, 1973) during which the addition of minute quantities of chitinase to *B. thuringiensis* preparations increased pathogenic activity against *Choristoneura fumiferana*. Their results showed that varieties *thuringiensis*, *finitimus*, *sotto*, *galleriae*, and *entomocidus* did not have any chitinolytic activity; *kurstaki* and *aizawai* had weak chitinolytic activity; and *alesti* and *dendrolimus* had high activity. The relationship between these properties and their pathogenicity are now known.

Finally, Pendleton et al. (1973) reported that two distinct hemolytic proteins, one similar to streptolysin O in its properties and behavior, are produced by a strain of *B. thuringiensis* var. *thuringiensis*. Thuringiolysin, so-called, is very similar to cereolysin from *B. cereus*, which is antigenically related to streptolysin O from group A streptococci. The approximate molecular weight of thuringiolysin is 47,000. Pronase rapidly inactivated hemolytic activity of the secondary hemolysin, and which therefore appears to be protein and seems not to be enzymatic, at least in respect to lipase, phospholipase, and esterase, which may be relevant to hemolytic activity of proteins. Again, the role that these hemolysins play in invertebrate infections is not known.

V. MODE OF ACTION

The most interesting characteristic of *B. thuringiensis* is that infection is not required for kill of certain of its insect hosts. Ingestion of the proteinaceous crystal by susceptible lepidopteran larvae kills the insects. The usual course of infection -- reproduction of bacterial cells within the host tissues, production of invasive factors by the bacillus, and toxin production within the host -- is not needed to produce toxemia. With other insects, the crystalline protein alone is not sufficiently toxic to cause quick kill, and the normal course of tissue invasion by viable cells takes place and does the killing. It is the ability of the parasporal crystal alone to produce kill (or the similar ability of several other toxins produced by this bacillus) that places the material in the realm of true toxins.

Other materials important to pathogenicity include the spores, which may or may not produce disease symptoms but do germinate in the insect gut. This germination gives rise to vegetative bacterial cells that, in turn, may produce exoenzymes such as phospholipases (lecithinases).

B. thuringiensis δ-exotoxin. Phospholipase (lecithinase) enzymes produced by crystalliferous and noncrystalliferous bacteria have been proposed as having some role in the insect mortality caused by the bacteria (Bonnefoi and Beguin, 1959; Heimpel, 1955a,b; Kushner and Heimpel, 1957; Toumanoff, 1953). The phospholipases catalyze reactions in which phospholipids are cleaved at various positions on the molecule, each particular enzyme (phospholipases A, C, and D and lysolecithinase) is specific for a particular site of action. Since, histologically, damage to the midgut cells of infected larvae involves parts of cells that include phospholipids, lecithinase or other phospholipase activity could be involved in the observed pathology. For example, phospholipase-catalyzed reactions probably lead to the disruption of the gut wall (observed dissolution of phospholipid material), thereby permitting invasion of the hemocoel by spore or vegetative forms of the bacteria. This invasion then can result in a typical bacteremia.

The importance of the phospholipases in the pathology produced by certain serotypes of *B. thuringiensis*, particularly to their use as toxicants, is likely minor because there is little evidence that there is any relationship between phospholipase production in a given serotype and its effectiveness as a toxicant for lepidopteran insects. Also, phospholipases must be produced by vegetative cells of the invading pathogen. For most of the insect hosts against which *B. thuringiensis* is active (parasporal crystal toxin), death occurs within a few hours after ingestion of the toxic crystal protein; vegetative cell factors are not required for kill, and the destruction of the midgut epithelium is the result of the crystal toxin alone. Any phospholipase effects would be additive to the massive damage done by the parasporal crystal. Further, the midgut pH of most susceptible Lepidoptera is higher than the pH values reported as optimum for the phospholipases. Even in the few Lepidoptera in which spores are known to germinate in the midgut (with active vegetative cell multiplication), the gut contents do not leak into the blood, which probably indicates absence of phospholipase activity (Heimpel, 1963).

B. thuringiensis β-exotoxin. The mode of action of the *B. thuringiensis* β-exotoxin was studied by Sebesta and Horska (1968) and Sebesta et al. (1969b). First, they studied the effects of injected β-exotoxin on the synthesis of desoxyribonucleic acid (DNA) ribonucleic acid (RNA), and protein in mice and found that a dose 10 times that required for an LD_{50} had no effect on the incorporation of ^{14}C-labeled thymidine into DNA or of ^{14}C-labeled cytidine into liver RNA was inhibited when the mice were given 1/3 the dose required for an LD_{50}. They concluded that the synthesis of DNA was not affected, that the β-exotoxin inhibited RNA synthesis, and that the site of action of the toxin lay between cytidine triphosphate (CTP) and RNA. Subsequently, they used *Escherichia coli* and was inhibited and that the inhibition was dependent on the molar relationship between adenosine triphosphate (ATP) and the β-exotoxin because the polymerase apparently prefers the exotoxin to ATP. Also, inhibition occurred regardless of whether the β-exotoxin was added before or after onset of polymerization.

The molecule exhibits differential inhibition of vegetative and sporulation-specific RNA polymerases from its host cell, *B. thuringiensis* (Johnson et al., 1975). In fact, it is a positive modifier of *B. thuringiensis* RNA polymerase *in vitro* and may function during sporulation. Whether its regulating effect is directed against RNA polymerase sigma factor or on certain promotor sites within the genome is not known. The β-exotoxin also has been reported to be a competitive inhibitor of adenyl cyclase (Grahme-Smith et al., 1975) and to prevent protein synthesis at the translation level (Somerville and Swain, 1975). Finally, when the β-exotoxin was injected into the yellow mealworm, *Tenebrio molitor*, RNA synthesis was inhibited. It thus appeared to inhibit the polymerization step involved with RNA, and acted as a toxic metabolite because it inhibited the nucleotidases and DNA-dependent RNA polymerase involved with ATP. Such postulated effects at the cellular level could result in the death of larvae and/or pupae of some Diptera and Lepidoptera, primarily by prevention of completion of pupation.

B. thuringiensis γ-exotoxin. This unidentified enzyme(s) is responsible for cleaning egg yolk agar and is not well understood. There is no proof of toxicity, and it is doubtful that the substance should be recognized as a toxin involved with pathogenicity.

B. thuringiensis δ-endotoxin. Although the δ-endotoxin activity appears to reside in both spores and crystals, insufficient evidence is available to conclude that the activities are identical. However, in this discussion, the δ-endotoxin will be equated with the insecticidal activity residing in the parasporal crystal.

As noted, the *B. thuringiensis* δ-endotoxin is generally highly selective for lepidopteran larvae but some strains are selective for dipteran and coleopteran larvae. Effectiveness varies between serotypes of *B. thuringiensis* and across the range of host species for a given δ-endotoxin-producing strain. Apparently, the same symptoms are exhibited by each susceptible species as the dose is changed, but the dose required to produce such effects as gut paralysis varies between insect species (Galowalia et al., 1973).

The following hypotheses have been offered to explain the molecular mode of action of the δ-endotoxin: (1) interference with midgut cell metabolic enzyme functions or membrane proteins in the midgut (Faust, 1968); (2) as a neurotoxin (Cooksey et al., 1969); (3) upset of transport processes across the midgut membrane (Fast and Donaghue, 1971; Fast and Morrison, 1972; Nishiitsutsuji-Uwo et al., 1979); (4) as an inhibitor of acetylcholinesterase (Koenigstedt and Groth, 1972); (5) initiator of self-digestion of the gut epithelium (Lüthy, 1973); (6) action as a phospholipase (Somerville and Pockett, 1976); and (7) as a respiratory uncoupler in midgut mitochondria (Faust et al., 1974b; Faust, 1975; Travers et al., 1976a,b; Faust and Travers, 1979). The biochemical and histological temporal sequence of *B. thuringiensis* δ-endotoxin intoxication in certain lepidopteran larvae that has been reported by investigators is summarized in Table 5.

In the alkaline gut of a susceptible insect such as *B. mori* the toxic component is released by action of the nonenzymic alkaline components on the crystalline structure and is then acted upon by gut enzymes (Faust et al., 1967). The toxin action after dissolution of the parasporal crystal body in the gut results in a paralysis of the gut within minutes after the crystals are ingested (Hannay, 1953; Angus, 1954; Young and Fitz-James, 1959). The development of paralysis is accompanied by a progressive increase in the alkalinity of the blood, which is thought to occur because the gut epithelium is so damaged that equilibration occurs between the highly buffered midgut contents (pH 10.2 to 10.5)

TABLE 5. Corrleation of biochemical and histological temporal sequence of *Bacillus thuringiensis* δ-endotoxin intoxication in certain lepidopteran larvae.

Time (min)	Biochemical Effects	Histological Effects
0-1	Degradation of *B.t.* δ-endotoxin. a/,b/ Glucose uptake stimulated but amino acid, carbonate ion, mono- and divalent cation transport unchanged. c/, d/ Acceleration O_2 uptake in H^+ transport system. e/, f/	---
1-5	Maximum glucose uptake. c/, d/ Maximum acceleration O_2 uptake, uncoupling of oxidative phosphorylation and cessation ATP production. e/, f/	---
5-10	Cessation glucose uptake, no changes in K^+, Na^+, Ca^+, Mg^+ in gut tissue. c/, d/	Irregular swelling and distortion of gut microvilli. g/
10-20	General breakdown permeability control; K^+ and other ions increase in hemolymph loss of osmotic integrity. d/, h/, i/ Rise in silkworm blood pH, initial paralysis of gut. j/	Epithelial gut cells swell, protrusions develop, endoplasmic reticulum disrupts; mitochondria increase in size, cristae disintegrate. g/, h/, l/, r/, s/
After 20	Paralysis of silkworm gut. m/, n/, o/	Epithelial cells disrupt, lyse and slough off into lumen. j/, k/, l/ p/, q/, t/

a/ Fast and Videnova, 1974. b/ Fast, 1975. c/ Fast and Donaghue, 1971. d/ Fast and Morrison, 1972. e/ Faust et al., 1974b. f/ Travers et al., 1976a,b. g/ Ebersold et al., 1977. h/ Fast and Angus, 1965. i/ Louloudes and Heimpel, 1969. j/ Heimbel and Angus, 1959. k/ Sutter and Raun, 1967. l/ Angus, 1970. m/ Hannay, 1953. n/ Angus, 1954. o/ Young and Fitz-James, 1959. p/ Angus and Heimpel, 1959. q/ Hoopingarner and Materu, 1964. r/ Nishiitsutusji-Uwo et al., 1979. s/ Endo and Nishiitsutsuji-Uwo, 1979. t/ Iizuka, 1974.

and the relatively poorly buffered hemolymph (pH 6.8). In fact, if the hemolymph of an uninfected silkworm larva is made alkaline (pH 8) by the addition of sterile nontoxic buffers, a general paralysis occurs that is indistinguishable from the paralysis cuased by the *B. thuringiensis* δ-endotoxin after ingestion of the toxin or crystal (Heimpel and Angus, 1959). Thus, the general paralysis seems to be a result of increased hemolymph alkalinity rather than a direct action of the toxin; and the increased alkalinity develops because of leakage across the deteriorated gut epithelium. Larvae of *B. mori, Antheraea pernyi,* and *Manduca sexta* later develop a general and fatal paralysis of the whole body.

Biochemically, glucose uptake is stimulated in the larval midgut within 1 min after the susceptible insect ingests the toxic crystals (Fast and Donaghue, 1971; Fast and Morrison, 1972; Murphy et al., 1976). Stimulation is maximum within 5 min, and glucose uptake ceases in 10 min. Control of cell permeability breaks down within another 5 min as levels of K^+ and other ions increase in the hemolymph (Fast and Morrison, 1972; Fast and Angus, 1965; Louloudes and Heimpel, 1969). However, K^+, Na^+, Ca^+, and Mg^{2+} do not accumulate in gut tissue during the first 10 min after toxin is given per os, and no change is detected in the rate at which K^+ turns over in the gut tissue. Ramakrishnan (1968) found the treated *B. mori* larvae had five times as much K^+ as healthy larvae in the hemolymph approximatley 90 min after ingestion of a lethal dose of *B. thuringiensis* δ-endotoxin. Also, Pendleton (1970) showed that K^+ levels in the hemolymph of larvae of *Samia cynthia ricini* began increasing 10 min after the insect ingested the δ-endotoxin and increased to more than twice normal values within 60 min; thereafter the level remained constant, even though parlysis did not occur until 140 min after dosing.

Histologically in lepidopterans, after 5 min of exposure the microvilli containing mitochondria swell rapidly and after approximately 10 min the fine structure of the epithelial cells begins to disrupt (Sutter and Raun, 1967; Ebersold et al., 1977). The contents of the mitochondria seem to dissolve during the period of continuous enlargement, leaving spherical cavities and the cristae of mitochondria disintegrate and dissolve, leaving an empty shell that continues to expand until it finally ruptures. The Golgi complexes also show structural changes as does the endoplasmic reticulum, which loses its ribosomes (Ebersold et al., 1977, 1978). Thus, the δ-endotoxin causes drastic changes to the cell organelles, but it is now known whether it acts exclusively at the surface of the gut epithelium or has to penetrate into the cells to cause the internal organelle damage. After 20-25 min the columnar cells "balloon" or extrude noticeably and severely disrupt the gut wall as they burst (Heimpel and

Angus, 1959; Angus and Heimpel, 1959; Hoopingarner and Materu, 1964; Fast and Angus, 1965; Sutter and Raun, 1967; Angus, 1970; Iizuka, 1974). In mosquito larvae, such as *Aedes aegypti*, similar histological disruption occurs when the larvae are dosed with *B. thuringiensis* var. *israelensis* (deBarjac, 1978b).

Fast and Morrison (1972) suggested that the function of the toxin may be to disrupt K^+ regulation. However, when Fast and Donaghue (1971) reported that glucose uptake of *B. mori* larval midgut was stimulated within 1 min after ingestion of crystals, they suggested that these cation regulation effects could be a secondary phenomenon. Fast and Morrison (1972) therefore undertook to determine if the primary action of the δ-endotoxin is to disrupt ion regulation in *B. mori* larval midgut tissue. They reasoned that if the *B. thuringiensis* δ-endotoxin acts in a similar manner its effects should be reflected in altered levels of K^+ in the gut tissue and/or the hemolymph of treated insects. The data showed that the K^+ concentration in gut tissue did not change significantly within the first 10 min after ingestion. They also showed that by 20 min the K^+ concentration had increased significantly and then continued ot increase until, or perhaps even after, the insects were paralyzed. Thus the δ-endotoxin apparently does not function primarily as a potassium ionophore, i.e., the changes in ion levels in hemolymph observed in their study and in those of Ramakrishnan (1968) and Pendleton (1970) appear to be secondary effects resulting from a breakdown of gut metabolism that occurs between 10 and 15 min after the δ-endotoxin is ingested.

To determine whether transport of other ions was a primary mechanism of the *B. thuringiensis* δ-endotoxin action, Fast and Morrison (1972) measured levels of Na^+, Ca^{++}, Mg^{++}, and Mn^{++} after administering the toxin per os. When analysis of variance and Tukey's test were applied, no statistically significant differences were found with time in the levels of either Ca^{++} or Mg^{++}. Also the significant differences in Na^+ concentration did not follow any particular trend and appeared to result from inherent variability of Na^+ concentration; Mn^{++} was not detected. Again, these findings indicate that changes in ion levels in hemolymph 10 min after treatment and therefore ion transport in susceptible insects, are secondary effects.

Apparently, the initial effect of the toxin is concerned in some manner with a mechanism that influences uptake of glucose. It then follows that the histological effects, that is, "ballooning" or extrusion of the columnar cells noted previously are also secondary effects of some initial biochemical upset.

Faust et al. (1974a) and Travers et al. (1967a,b) tested degradation products (molecular weights 230,000, 67,000, 30,000, ~5000, and ~1000) of the parasporal crystals of *B. thuringiensis* var. *kurstaki* obtained when the crystals were treated with midgut juice from larvae of *B. mori* against midgut mitochondria preparations of *B. mori*. Manometric and colorimetric techniques revealed that the lower molecular weight polypeptides stimulated the oxygen uptake of midgut mitochondria and inhibited ATP production. Thus the *B. thuringiensis* δ-endotoxin may act as an uncoupler of oxidative phosphorylation. The loss of ATP production caused by this action would lead to metabolic imbalance and perhaps cell death.

Faust et al. (1974b) and Travers et al. (1976b) therefore offered a theory whereby they integrated the histological and biochemical evidence of the molecular mode of action. Since the *B. thuringiensis* δ-endotoxin increases oxygen uptake by uncoupling oxidative phosphorylation in mitochondria and results in a low level of ATP production, the high demand for reduced coenzymes necessitated by increased oxygen uptake in a non-conservative electron transport system would at first increase the demand of the insect's catabolic process for glucose. Such stimulation of respiration by the *B. thuringiensis* δ-endotoxin might explain the stimulation of glucose uptake during the first 5 min of intoxication in susceptible insects. Meanwhile, in the cell the ratio of the diphosphate ester ADP and the triphosphate ester ATP controls the rate at which NADH, the major reduced coenzyme in the cell, is used in the processes of electron transport. When ATP levels fall because of the action of the uncoupler, the rate of NADH use increases dramatically. In fact, La Nove et al. (1970) reported that levels of cellular NADH are the major factor controlling the Krebs citric acid cycle, and increased NADH utilization in an uncoupled electron transport system would cause stimulation of the TCA cycle. This increased TCA cycle activity would, in turn, cause stimulation of glycolysis and a concurrent increase in glucose uptake as observed by Fast and Donaghue (1971), that is, high demand for reduced biological intermediates by increased oxygen uptake in a non-conservative electron transport system would increase the demand in the insect's catabolic process for glucose. This increased demand could explain the sudden influx, noted by Fast and Donaghue (1971), of glucose into the midgut cells of the insect. This interpretation was supported by a rapid (half-time 2 min) depletion of ATP caused by active toxin observed in cultured cells susceptible to the toxin (Fast, pers. commun.). Such a depletion of ATP would stimulate respiration and subsequent uptake of glucose from the medium. However, the ATP loss observed was not simply the result of a shortage of glucose in the medium because tripling the glucose concentration did not alter the rate of ATP loss in cultured cells.

Then, by 10 min, glucose uptake ceases, probably because respiratory chain-linked phosphorylation is ceasing [analysis by Travers et al. (1976a) of inorganic phosphate revealed that ATP production was being inhibited in the affected systems]. As a result the energy for glucose metabolism and other energy-requiring activities becomes limited. The initial stimulation of glucose flux could therefore result from the initial effect of the δ-endotoxin on the metabolic respiratory system, that is, glucose uptake increases as metabolic respiration increases, but no longer has the benefit of respiratory chain-linked ATP production. As endogenous ATP becomes limited, glucose uptake stimulated by increased respiration ceases and the changes that have been observed between 15 and 25 min should follow -- ion transport upset, "ballooning," bursting from uncontrolled osmotic effects, and disintegration of the midgut wall. [In insects, absorption of glucose occurs by a diffusion process across the gut wall rather than by some active transporting system of the type believed to occur in sugar absorption across the mammalian gut wall (Treherne 1958a,b).]

Faust et al. (1974b) demonstrated that toxic fragments of the *B. thuringiensis* δ-endotoxin increased in activity against the mitochondrial preparations as fragment size decreased down to approximately 30,000 daltons. Fragments of 230,000 daltons had no significant effect.

The theory of uncoupling of mitochondrial ATP synthesis is compatible with the valinomycin toxicity data of Angus (1968b) since valinomycin is also an uncoupling agent (McMurray and Begg, 1959). When Angus (1968b) tested valinomycin against 5th-instar *B. mori* larvae he found that after 45 to 60 min the larvae ceased feeding, after 60 to 90 min the insects became sluggish, and after 90 to 180 min the insects became paralyzed and the hemolymph alkalinity increased to pH 7.8 to 8.0. (Changes in glucose uptake and ion imbalance apparently were not determined). Angus hypothesized that since the permeability of thin lipid membranes (Andreoli et al., 1967) and of sheep red blood cells (Toteson et al., 1967) is affected by valinomycin, the δ-endotoxin acts on the membrane of the midgut epithelial cells and affects their selective permeability. However, this conclusion is rather speculative because one might also expect the known effects of valinomycin on mitochondria to cause the secondary effects reported by Angus (1968b). Indeed, the effect of valinomycin on mitochondria was first reported by McMurray and Begg (1969), who showed that the compound uncoupled respiration from phosphorylation and stimulated ATP activity.

Nishiitsutsuji-Uwo et al. (1979) compared the cytological effects of valinomycin and δ-endotoxin by using TN-368 cell lines from *Trichoplusia ni*. Mitochondria of the toxin-treated cells were transformed into a form in which the matrix was very "condensed" and extraordinarily electron dense; those of valinomycin-treated cells were transformed into the "swollen" form, i.e., they showed matrical swelling, the development of a markedly swollen, balloon like appearance with very low electron density. The cristae in the swollen mitochondria were narrow and tubular. Also, they found ultrastructural differences in the mitochondria that suggested that the function of δ-endotoxin differs somewhat from that of valinomycin. This is not surprising since the structures of the two proteins are dissimilar.

Other researchers (Fast et al., 1978) favor a direct interaction of the δ-endotoxin with the cell membrane of the gut epithelium. They showed that enzymatically activated δ-endotoxin of *B. thuringiensis* covalently bound to Sephadex beads had the same effect on insect cells in tissue culture (CF-124 cells from *Choristoneura fumiferana*) as free toxin by measuring residual ATP. The effect was prevented by antitoxin-antibody and heat denaturation and apparently was not produced by a nonspecific protein effect or by the beads. They thus concluded that the toxin probably acts at the cell surface but did not determine whether bound toxin was released by extracellular proteolytic digestion in the tissue culture system. They did, however, determine that the observed effect was caused by toxin that was resolubilized when they incubated bound toxin for 30 min in the absence of cells and then assayed the bead-free incubate against cells of the CF-124 line. Some reduction in cellular ATP resulted from exposure to activated δ-endotoxin, but it was not sufficient to account for the results obtained with the bound toxin. If the initial effect of the δ-endotoxin was on the cell surface membrane, then cation flux presumably would precede (cause and effect) any increase in glucose uptake. The fact that glucose uptake is initially stimulate in 1 min and ends within 10 min seems to agree with the effects of uncoupling oxidative phosphorylation from the electron transport system.

Interestingly, Yousten and Guerrant (1976) found no evidence of any similarity of action of *B. thuringiensis* δ-endotoxin, cholera toxin, and *Escherichia coli* enterotoxin when they were tested against cabbage looper, rabbit ileal loop secretion, and rabbit intestine and no demonstrable antigenic relationship between *B. thuringiensis* δ-endotoxin and cholera toxin. All three bacterial toxins induce marked fluid accumulation in the intestines of their respective hosts, but cholera toxin and enterotoxin appeared to activate adenylate cylase in mammalian intestine and in many other cell types. Also, cholera toxin binds in a first step to the cell membrane of the intestine.

Cooksey et al. (1969) examined sheathed and desheathed cercal nerve and abdominal ganglia 5 and 6 that had been isolated from adult cockroaches (Orthoptera) and reported that nerve preparations exposed to hydrolyzed *B. thuringiensis* δ-endotoxin for 20 min retained full sensitivity to acetylcholine; thus the toxin was probably not acting at a postsynaptic site. Also hydrolyzed *B. thuringiensis* δ-endotoxin had no significant effect on axonal conduction when it was applied for a period of 12 hr to a giant fiber axonal preparations. In the opinion of Cooksey et al. (1969), the absence of effect confirmed the data of Angus (1968b) and Ramakrishnan (1968), i.e., the δ-endotoxin did not act as an ionophore. The results of Cooksey et al. (1969) did demonstrate that the δ-endotoxin was unable to affect the axon, whether sheathed or desheathed, when the toxin was placed on the axon alone. If the δ-endotoxin was a surface active agent, as has been postulated, desheathed nerve axons would be more susceptible. It appears that the δ-endotoxin is unable to diminish the ion gradient across the axon in these experiments.

No data have yet been presented as to the character of any toxic moieties that may enter the midgut epithelial cells. Much more work needs to be done in these areas to definitely determine the initial site(s) of action of the *B. thuringiensis* δ-endotoxin. All one can conclude from the reported evidence is that the toxin acts at the cell surface or within the cell to cause a rapid loss of ATP from those cells that stimulate respiration and glucose uptake and that stimulation of glucose uptake in the insect *in vivo* reaches a maximum about 5 min after dosing, about the time cellular ATP has been depleted in cultured cells and ATP production has ceased in isolated mitochondria. This effect is followed shortly by histological changes in the microvilli, mitochondria, and other cell organelles.

Other toxins. The modes of action of the labile exotoxin (Smirnoff and Berlinguet, 1966), water-soluble toxin (Fast, 1971), thermo-sensitive exotoxin (Krieg, 1971), and thuringiolysin (Pendleton et al., 1973) have not been determined. The relationship between the chitinolytic enzyme properties of certain *B. thuringiensis* strains and their pathogenicity also is unknown.

VI. CONCLUSION

B. thuringiensis is one of the most versatile pathogens yet found in insect control research. It kills numerous species of insects, notably caterpillars, flies, and mosquitoes, that are economic problems, yet ordinarily does not harm beneficial insects or other non-target forms of life. The toxic manifestations induced by the ingestion of *B. thuringiensis* δ-endotoxin are anorexia in the first few hours, followed in most cases by lethargy, general paralysis, and changes in the gut leading to death. These toxic effects, due to the parasporal crystal toxin alone, may be accompanied by septicemia, when spores are present or when the insect has ingested old sporulated cultures. Thus, in the pathogenic process of *B. thuringiensis* one can distinguish between toxemia, brought on by the crystal alone, and other pathological manifestations related to the presence of spores or other toxic entities produced by the bacteria. When conditions are favorable for sporulation or vegetative growth in the gut, other toxins such as phospholipase C, which is secreted by some strains, may act to reinforce the toxemia and facilitate bacterial penetration into the hemolymph. The role of the *B. thuringiensis* β-exotoxin in natural infections is unknown. Its pathology, however, has been well established in laboratory tests of supernatant fluids from cultures of *B. thuringiensis*.

Natural epizootics in insects attributed to *B. thuringiensis* are infrequent and limited to septicemia in populations of a few species such as *Dendrolimus sibericus*, *Anagasta kuehniella*, *Selenephera lunigera*, and *Cadra cautella*. However, the isolation and identification of *B. thuringiensis* strains from naturally infected insects indicates that the organism persists in the natural environment and is possibly a natural enzootic regulatory factor. Apparently, *B. thuringiensis* usually does not reach the level required for an epizootic to develop in an insect population because spores (and crystals) can only be found in small numbers in the host's cadaver. Commercial interest in *B. thuringiensis* and its lethal agents are not due, therefore, to its power of contagion but to its selective killing of susceptible insects by primarily the β-exotoxin and the δ-endotoxin when applied as a microbial control agent.

VII. REFERENCES

Akune, S., Watanabe, T., Mukai, J., Tsutsui, R., and Abe, K. (1971). A toxic protein from the insect pathogen *Bacillus thuringiensis*. *Jap. J. Med. Sci. Biol.*, 24, 57-59.

Andreoli, T. E., Tieffenberg, M., and Toteson, D. C. (1967). The effect of valinomycin on the ionic permeability to thin liquid memberanes. *J. Gen. Physiol.*, 50, 2527-2545.

Angus, T. A. (1954). A bacterial toxin paralyzing silkworm larvae. *Nature*, 173, 545-546.

Angus, T. A. (1956b). Association of toxicity with protein-crystalline inclusions of *Bacillus sotto* Ishiwata. *Can. J. Microbiol.*, 2, 122-131.

Angus, T. A. (1956c). Extraction, purification, and properties of *Bacillus sotto* toxin. *Can. J. Microbiol.*, 2, 416-426.

Angus, T. A. (1968a). The use of *Bacillus thuringiensis* as a microbial insecticide. *World Rev. Pest Control*, 7, 11-26.

Angus, T. A. (1968b). Similarity of effect of valinomycin and *Bacillus thuringiensis* parasporal protein in larvae of *Bombyx mor*. *J. Invert. Pathol.*, 11, 145-146.

Angus, T. A. (1970). Implications of some recent studies of *Bacillus thuringiensis* - a personal preview. In *Proceedings IV International Colloquium on Insect Pathology*, College Park, Maryland. pp. 183-189.

Angus, T. A. and Heimpel, A. M. (1959). Inhibition of feeding and blood pH changes in Lepidopterous larvae infected with crystal-forming bacteria. *Can. Entomol.*, 91, 352-358.

Angus, T. A. and Norris, J. R. (1968). A comparison of the toxicity of some varieties of *Bacillus thuringiensis* Berliner for wilkworm larvae. *J. Invert. Pathol.*, 11, 289-295.

Aronson, J. N. and Tillinghast, J. (1976). A chemical study of the parasporal crystal of *Bacillus thuringiensis*. *Spore Res.*, 1, 351-357.

deBarjac, H. (1978a). Toxicite de *Bacillus thuringiensis* var. *israelensis* pour les larves d'*Aedes aegypti* et d'*Anopheles stephensi*. *C. R. Acad. Sci. Paris*, 286, 1175-1178.

deBarjac, H. (1978b). Etude cytologique de l'action de *Bacillus thuringiensis* var. *israelensis* sur larves de mousquitoes. *C. R. Acad. Sci. Paris*, 286, 1629-1632.

deBarjac, H. (1978c). Une novelle variete de *Bacillus thuringiensis* tres toxique pour les moustiques: *B. thuringiensis* var. *israelensis* serotype 14. *C. R. Acad. Sci. Paris*, 286, 797-800.

deBarjac, H. and Bonnefoi, A. (1962). Essai de classification biochimique et serologique de 24 souches de *Bacillus* du type *B. thuringiensis*. *Entomophaga*, 7, 5-31.

deBarjac, H. and Dedonder, R. (1965). Isolement d'un nucleotide identifiable a la "toxine thermostable" de *Bacillus thuringiensis* var. *Berliner*. *C. R. Acad. Sci. Paris*, 260, 7050-7053.

deBarjac, H., Burgerjon, A., and Bonnefoi, A. (1966). The production of heat-stable toxin by nine serotypes of *Bacillus thuringiensis*. *J. Invert. Pathol.*, 8, 537-538.

deBarjac, H. and Bonnefoi, A. (1967). Classification des souches de *Bacillus thuringiensis*. *C. R. Acad. Sci. Paris*, 264, 1811-1813.

deBarjac, H. and Bonnefoi, A. (1968). A classification of strains of *Bacillus thuringiensis* Berliner with a key to their differentiation. *J. Invert. Pathol.*, 11, 335-347.

deBarjac, H. and Dedonder, R. (1968). Purification de la toxine thermostable de *B. thuringiensis* var. *thuringiensis* et analysis complementaries. *Bull. Soc. Chim. Biol.*, 50, 941-944.

Bateson, J. B. and Stainsby, G. (1970). Analysis of the active principle in the biological insecticide *Bacillus thuringiensis* Berliner. *J. Food Technol.*, 5, 403-415.

Bechtel, D. B. and Bulla, L. A. (1976). Electron microscope study of sporulation and parasporal crystal formation in *Bacillus thuringiensis*. *J. Bacteriol.*, 127, 1472-1481.

Berliner, E. (1911). Uber die Schlaffsucht der Mehlmottenroupe. *Z. Ges. Getreidew.*, 3, 63-70.

Berliner, E. (1915). Uber die Schlaffsucht der Mehlmottenroupe (*Ephestia kuhniella* Zell.) und ihren Erreger *Bacillus thuringiensis*, n.sp. *Z. Angew. Entomol.*, 2, 29-56.

Bond, R. P., Bocye, C. B., and French, S. J. (1969). A purification and some properties of an insecticidal exotoxin from *Bacillus thuringiensis* Berliner. *Biochem. J.*, 114, 477-488.

Bond, R. P., Boyce, C. B., Rogoff, M. H., and Shiefh, T. R. (1971). The thermostable exotoxin of *Bacillus thuringiensis*. *In* "Microbial Control of Insects and Mites," H. D. Burges and N. W. Hussey (Eds.). Academic Press, New York. pp. 275-303.

Bonnefoi, A. and Beguin, S. (1959). Recherches sur l'action des cristaux de *Bacillus thruingiensis* souche "anduze". *Entomophaga.* 4, 193-199.

Bonnefoi, A. and deBarjac, H. (1963). Classification des souches du groupe *Bacillus thuringiensis* par la determination de l'antignee flagellaire. *Entomophaga.* 8, 223-229.

Bucher, G. (1959). The bacterium *Coccobacillus acridiorum* d'Herelle: Its taxonomic position and status as a pathogen of locusts and grasshoppers. *J. Insect Pathol.* 1, 331-346.

Bucher, G. E. (1963). Nonsporulating bacterial pathogens. *In* "Insect Pathology: An Advanced Treatise", E. A. Steinhaus (ed.). Vol. 2, Academic Press, New York. pp. 117-147.

Bucher, G. E. and Stephens, J. (1957). A disease of grasshoppers caused by the bacterium *Pseudomonas aeruginosa* (Schroeter) Migula. *Can. J. Microbiol.* 3, 611-625.

Bucher, G. E., Angus, T. A., and Krywienczyk, J. (1966). Characteristics of a new strain of *Bacillus thuringiensis* var. *thuringiensis* Berliner (Serotype I) isolated from the bumblebee wax moth. *J. Invert. Pathol.* 8, 485-491.

Bulla, L. A., Kramer, J. J., and Davidson, L. I. (1977). Characterization of the entomocidal parasporal crystal of *Bacillus thuringiensis*. *J. Bacteriol.* 130, 375-383.

Burgerjon, A. and deBarjac, H. (1960a). Essais preliminaries sur le role insecticide de la toxine thermostable produite par *Bacillus thuringiensis* Berliner. *Proceedings International Congress Entomology, 11th Vienna.* 2, 835-839.

Burgerjon, A. and deBarjac, H. (1960b). Nouvelle donnus sur le role de la toxine soluble thermostable produite par *Bacillus thuringiensis* Berliner. *C. R. Acad. Sci. Paris.* 251, 199-912.

Burgerjon, A. and deBarjac, H. (1964). Etude de la toxine soluble thermostable chez differentes souches de *Bacillus thuringiensis*. *Colloq. Int. Pathol. Insectes, Paris, 1962, Entomophaga, Mem.* 2, 221-226.

Burgerjon, A. and Biache, G. (1967). Divers effects speciaux et symptomes teratologiques de la toxine thermostable de *B. thuringiensis* en fonction de l'age physiologique des insectes. *Ann. Soc. Entomol. Fr., N.S.* 3, 929- 952.

Burgerjon, A., Biahce, G., and Cals, P. (1969). Terotology of the Colorado Potato Beetle, *Leptinotarsa decemlineata*, as provoked by larval administration of the thermostable toxin of *Bacillus thuringiensis*. *J. Invert. Pathol.* 14, 273-278.

Burges, H. D. and Hurst, J. A. (1977). Ecology of *Bacillus thuringiensis* in storage moths. *J. Invert. Pathol.* 30, 131-139.

Burges, H. and Hussey, N. (1971). *In* "Microbial Control of Insects and Mites", H. Burges and N. Hussey (eds.). Academic Press, New York.

Cantwell, G. E., Heimpel, A. M., and Thompson, M. J. (1964). The production of an exotoxin, the fly factor, by various crystal forming bacteria related to *Bacillus thuringiensis* var. *thuringiensis*. *J. Insect Pathol.* 6, 406-408.

Carlton, B. C. (1976). Complex plasmid system of *Bacillus megaterium*. *In* "Microbiology-76", American Society for Microbiology, Washington, D.C. pp. 394-405.

Chestukhina, G. G., Kostina, L. I., Zalunin, I. A., Kotova, T. S., Katrukha, S. P., Kuznetsov, Y. S., and Stepanov, V. M. (1977). Proteins of *Bacillus thuringiensis* δ-endotoxin crystals. *Biokhimiya.* 42, 1660-1667.

Chestukhina, G. G., Zalunin, I. A., Kostina, L. I., Kotova, T. S., Katrukha, S. P., Lyublinskaya, L. A., Stepanova, V. M. (1978). Proteinases bound to crystals of *Bacillus thuringiensis*. *Biokhimiya.* 43, 857-864.

Clowes, R. C. (1972). Molecular structure of bacterial plasmids. *Bacteriol. Rev.* 36, 361-405.

Cooksey, K. E. (1968). Purification of a protein from *Bacillus thuringiensis* toxic to larvae of lepidoptera. *Biochem. J.* 106, 445-454.

Cooksey, K. E., Donninger, C., Norris, J. R., and Shankland, D. (1969). Nerve blocking effect of *Bacillus thuringiensis* protein. *J. Invert. Pathol.* 13, 461-462.

Debabov, V. G., Azizbekyan, R. R., Chilabalina, O. I., Djarhenko, V. V., Galushka, F. P., and Belich, R. A. (1977). Isolation and preliminary characterization of extrachromosomal elements of *Bacillus thuringiensis* DNA. *Genetica* (USSR). 13, 496-500.

Delaporte, B. and Beguin, S. (1955). Etude d'une souche de *Bacillus* pathogene pour certains insects, identifiable a *Bacillus thuringiensis* Berliner. *Ann. Inst. Pasteur.* 89, 632-643.

Dulmage, H. T. (1970). Production of the spore δ-endotoxin complex by variants of *Bacillus thuringiensis* in two fermentation media. *J. Invert. Pathol.* 16, 385-389.

Dulmage, H. T. (1971). Production of δ-endotoxin by eighteen isolates of *Bacillus thuringiensis*, serotype 3, in three fermentation media. *J. Invert. Pathol.* 18, 353-358.

Dulmage, H. T. et al. (1981). Insecticidal activity of isolates of *Bacillus thuringiensis* and their potential for pest control. *In* "Microbial Control of Pests and Plant Diseases, 1970-1980". H. D. Burges (ed.). Academic Press, London.

Ebersold, H. R., Luthy, P., and Mueller, M. (1977). Changes in fine structure of the gut epithelium of *Pieris brassicae* induced by the δ-endotoxin of *Bacillus thuringiensis*. *Bull. Soc. Entomol.*, Suisse. 50, 269-276.

Ebersold, H. R., Lüthy, P., Geiser, P., and Ettlinger, L. (1978). The action of the δ-endotoxin of *Bacillus thuringiensis:* an electron microscope study. *Experientia.* 34, 1672.

Endo, Y. and Nishiitsutsuji-Uwo, J. (1979). Ultrastructural changes in the midgut epithelium of *Bombyx mori* induced by *Bacillus thuringiensis* crystals. *Jap. J. Appl. Entomol. Zool.* 23, 183-185.

Ermakova, L. M., Galushka, F. P., Strongin, A. Ya., Sladkova, I. A., Rebentish, B. A., Andreeva, M. V., and Stepanova, V. M. (1978). Plasmids of crystal-forming bacilli and the influence of growth medium composition on their appearance. *J. Gen. Microbiol.* 107, 169-171.

Farkas, J., Sebeska, K., Horska, K., Samek, Z., Dolejs, J., and Sorm, F. (1969). The structure of exotoxin of *Bacillus thuringiensis* var. *gelechia. Collect. Czech. Chem. Commun.* 34, 118-1119.

Fast, P. G. (1971). Isolation of a water-soluble toxin from a commercial microbial insecticide based on *Bacillus thuringiensis*. *J. Invert. Pathol.* 17, 301.

Fast, P. G. (1975). Purification of fragments of *Bacillus thuringiensis* δ-endotoxin from hemolymph of spruce budworm. *In "Can. For. Serv. Biomonthy Res. Notes".* 31, 1-2.

Fast, P. G. and Angus, T. A. (1965). Effects of parasporal inclusions of *Bacillus thuringiensis* var. *sotto* Ishiwata on the permeability of the gut wall of *Bombyx mori* (Linnaeus) larvae. *J. Invert. Pathol.* 7, 29-32.

Fast, P. G. and Angus, T. A. (1970). The δ-endotoxin of *Bacillus thuringiensis* var. *sotto:* a toxic low molecular weight fragment. *J. Invert. Pathol.* 16, 465.

Fast, P. and Donaghue, T. (1971). The δ-endotoxin of *Bacillus thuringiensis.* II. On the mode of action. *J. Invert. Pathol.* 18, 135-138.

Fast, P. and Morrison, I. (1972). The δ-endotoxin of *Bacillus thuringiensis.* IV. The effect of δ-endotoxin on ion regulation by midgut tissue of *Bombyx mori* larvae. *J. Invert. Pathol.* 20, 208-211.

Fast, P. and Videnova, E. (1974). The δ-endotoxin of *Bacillus thuringiensis.* V. On the occurrence of endotoxin fragments in hemolymph. *J. Invert. Pathol.* 18, 135-138.

Faust, R. M. (1968). In vitro chemical reaction of the δ-endotoxin produced by *Bacillus thuringiensis* var. *dendrolimus* with other proteins. *J. Invert. Pathol.* 11, 465-475.

Faust, R. M. (1975). Toxins of *Bacillus thuringiensis:* Mode of action. *In* "Biological Regulation of Vectors", J. D. Briggs (ed.). HEW, NIH, NIAID, Pub. 77-1180. pp. 31-48.

Faust, R. M., Hallam, G. M., and Travers, R. S. (1973). Spectrographic elemental analysis of the parasporal crystals produced by *Bacillus thuringiensis* var. *dendrolimus* and the polyhedral inclusion bodies of the nucleopolyhedrosis virus of the fall armyworm, *Spodoptera frugiperda. J. Invert. Pathol.* 22, 478-480.

Faust, R., Hallam, G, and Travers, R. (1974a). Degradation of the parasporal crystal produced by *Bacillus thuringiensis* var. *kurstaki. J. Invert. Pathol.* 24, 365-373.

Faust, R., Travers, R., and Hallam, G. (1974b). Preliminary investigations on the molecular mode of action of the δ-endotoxin produced by *Bacillus thuringiensis* var. *alesti*. *J. Invert. Pathol.* 23, 259-261.

Faust, R. M., Spizizen, J., Gage, V., and Travers, R. S. (1979). Extrachromosomal DNA in *Bacillus thuringiensis* var. *kurstaki*, var. *finitimus*, var. *sotto*, and in *Bacillus popilliae*. *J. Invert. Pathol.* 33, 233-238.

Faust, R. M. and Travers, R. S. (1979). Effects of *Bacillus thuringiensis* δ-endotoxin on the midgut cytochrome system of *Bombyx mori*. *In* "Abstracts Annual Meeting, American Society for Microbiology", Washington, D.C. p. 31.

Fujiyoshi, N. (1973). Studies on the utilization of spore-forming bacteria for the control of house flies and mosquitoes. *In* "Research Report of the Seibu Chemical Industry Co. Ltd", Special Issue No. 1. pp. 37.

Galowalia, M. M. S., Gibson, N. H. E., and Wolf, J. (1973). The comparative potencies of the crystalline endotoxin of eight varieties of *Bacillus thuringiensis* to larvae of *Pieris brassicae*. *J. Invert. Pathol.* 21, 301-308.

Galushka, F. P. and Azizbekyan, R. R. (1977). Investigations of plasmids of strains of different variants of *Bacillus thuringiensis*. *Dokl. Akad. Nauk. SSR.* 236, 1233-1235.

Galtron, M. F., Lecadet, M. M., and Dedonder, R. (1972). Structure of the parasporal inclusion of *Bacillus thuringiensis* Berliner: Characterization of a repetitive subunit. *Eur. J. Biochem.* 30, 330-338.

Grahme-Smith, D. G., Isaac, P., Heal, D. J., and Bond, R. P. M. (1975). Inhibition of adenyl cyclase by an exotoxin of *Bacillus thuringiensis*. *Nature.* 253, 58-60.

Grigorova, R. (1964). Deux souches de *Bacillus thuringiensis* Berliner isolees des chenilles de *Bombyx disparate* et *Lymantria dispar*. *Colloquium International Pathology Insectes, Paris, 1962 Entomophage Mem.* 2, 179-191.

Grigorova, R. E., Kantardgieva, E., and Pashov, N. (1967). On the shape and structure of the crystal in two strains of *Bacillus thuringiensis*. *J. Invert. Pathol.* 9, 503-509.

Gukasyan, A. B. (1961). Prospective control method for the Siberian silkworm. *Akad. Nauk SSR Vestn.* 35, 58-59.

Hall, I. M., Arakawa, K. Y., Dulmage, H. T., and Correa, J. A. (1977). The pathogenicity of strains of *Bacillus thuringiensis* to larvae of *Aedes* and to *Culex* mosquitoes. *Mosquito News*, 37, 246-251.

Hannay, C. L. (1953). Crystalline inclusions in aerobic spore-forming bacteria. *Nature*, 172, 1004.

Hannay, C. (1956). Inclusions in bacteria. *In* "Bacterial Anatomy," E. Spooner and B. Stocker (eds.). pp. 318-340.

Hannay, C. L. and Fitz-James, P. (1955). The protein crystals of *Bacillus thuringiensis* Berliner. *Can. J. Microbiol.*, 1, 694-710.

Heimpel, A. M. (1954a). Investigations of the mode of action of strains of *Bacillus cereus* Frankland and Frankland pathogenic for the larch sawfly, *Pristiphora erichsonii* (Htg.) and other insects with reference to the pathogenicity of *Bacillus cereus* Fr. and Fr. *Ph.D. Thesis*, Queen's Univ., Kingston, Canada. 155 pp.

Heimpel, A. M. (1954b). A strain of *Bacillus cereus* Fr. and Fr. pathogenic for the larch sawfly, *Pristiphora erichsonii* (Htg.) *Can. Entomol.*, 86, 73-77.

Heimpel, A. M. (1955a). The pH in the gut and blood of the larch sawfly, *Pristiphora erichsonni* (Htg.), and other insects with reference to the pathogenicity. *Can. J. Zool.*, 33, 99-106.

Heimpel, A. M. (1955b). Investigations of the mode of action of strains of *Bacillus cereus* Frankland and Frankland pathogenic for the larch sawfly, *Pristiphora erichsonni*. *Can. J. Zool.*, 33, 311-326.

Heimpel, A. M. (1963). The status of *Bacillus thuringiensis*. *Advan. Chem. Ser.*, No. 41, pp. 64-74.

Heimpel, A. M. (1965). The specificity of the pathogen *Bacillus thuringiensis* var. *thuringiensis* Berliner for insects. *Proc. 12th International Congress Entomology*, London, 1964. p. 736.

Heimpel, A. M. (1967). A critical review of *Bacillus thuringiensis* var. *thuringiensis* Berliner and other crystalliferous bacteria. *Annu. Rev. Entomol.*, 12, 287-322.

Heimpel, A. M. and Angus, T. A. (1958a). Recent advances in the knowledge of some bacterial pathogens of insects. *Proceedings 10th International Congress Entomology*, 4, 711-712.

Heimpel, A. M. and Angus, T. A. (1958b). The taxonomy of insect pathogens related to *Bacillus cereus* Fr. and Fr. *Can. J. Microbiol.*, 4, 531-541.

Heimpel, A. M. and Angus, T. A. (1959). The site of action of crystalliferous bacteria in Lepidoptera larvae. *J. Insect Pathol.*, 1, 152-170.

Heimpel, A. M. and Angus, T. (1963). Diseases caused by certain spore-forming Bacteria. *In* "Insect Pathology: An Advanced Treatise," Vol. 1, E. A. Steinhaus (ed.). pp. 21-73. Academic Press, New York.

Herbert, B. N., Could, H. J., and Chain, E. B. (1971). Crystal protein of *Bacillus thuringiensis* var. *tolworthi:* subunit structure and toxicity to *Pieris brassicae*. *Eur. J. Biochem.*, 24, 366-375.

Herbert, B. N. and Gould H. J. (1973). Biosynthesis of the crystal protein of *Bacillus thuringiensis* var. *tolworthi*. 1. Kinetics of formation of the polypeptide components of the crystal protein *in vivo*. *Eur. J. Biochem.*, 37, 441-448.

d'Herelle, F. (1914). Le coccobacille des sauterelles. *Ann. Inst. Pasteur.*, 28, 280-328, 387-407.

Holmes, K. C. and Monro, R. E. (1965). Studies on the structure of parasporal inclusions from *Bacillus thuringiensis*. *J. Mol. Biol.*, 14, 572-581.

Hoopingarner, R. and M. E. Materu. (1964). Toxicology and histopathology of *Bacillus thuringiensis* Berliner in *Galleria mellonella* (L.). *J. Insect Pathol.*, 6, 26-39.

Husz, B. (1928). *Bacillus thuringiensis* Berliner, a bacterium pathogenic to corn borer larvae, a preliminary report. *Sci. Rep. Int. Corn Borer Invest.*, 1, 191-193.

Husz, B. (1930). Field experiments on the application of *Bacillus thuringiensis* against the corn borer. *Sci. Rep. Int. Corn Borer Invest.*, 3, 91-93.

Husz, B. (1931). Experiments during 1931 on the use of *Bacillus thuringiensis*. Berliner in controlling the corn borer. *Sci. Rep. Int. Corn Borer Invest.*, 4, 22-23.

Iizuka, T. (1974). Histo- and cyto- pathological studies on the midgut epithelium of silkworm larvae fed *Bacillus thuringiensis* *J. Fac. Agric. Hokkaido Univ.*, 57, 313-318.

Iizuka, T., Faust, R. M., and Travers, R. S. (1981a). Isolation and partial characterization of extrachromosomal DNA from serotypes of *Bacillus thuringiensis* pathogenic to lepidopteran and dipteran larvae by agarose gel electrophoresis. *J. Sericult. Sci. Japan.*, 50, 1-14.

Iizuka, T., Faust, R. M., and Travers, R. S. (1981b). Comparative profiles of extrachromosomal DNA in single and multiple crystalliferous strains of *Bacillus thuringiensis* variety *kurstaki*. *J. Fac. Agr. Hokkaido Univ.*, 60, 143-151.

Ishiwata, S. (1901). On a kind of severe flacherie (sotto disease) (I) *Dainilon Sanshi Kaiho*, 9, 1-5. (In Japanese).

Ivinskene, V. L., Fluer, F. S., and Vasilyauskas, I. F. (1975). Formation of lecithinase by *Bacillus thuringiensis*. *Mikrobiologiya*, 44, 999-1004. (In Russian).

Jacobs, S. E. (1950). Bacteriological control of the flour moth *(Ephestia kühniella)*. *Proc. Soc. Appl. Bacteriol.*, 13, 83-91.

Johnson, C. E., Bulla, L. A., Jr., and Nickerson, K. W. (1975). Differential inhibition of β-exotoxin of vegetative - and sporulation - specific ribonucleic and polymerases from *Bacillus thuringiensis* cells. *In* "Spores VI," P. Gerhardt, R. N. Costilow, and H. L. Sadoff, (eds.), *American Society Microbiology*, Washington, D.C. pp. 248-254.

Koenigstedt, D. and Groth, V. (1972). Zur Wirkung des endotoxins von *Bacillus thuringiensis* Berliner auf larven von *Pieris brassicae*. *Biol. Rundsch.*, 10, 389-394.

Krieg, A. (1961). *Bacillus thuringiensis* Berliner. *Mittbiol. Bundesanst. Laud Forstwistsch. Berlin-Dahlem.*, 103, 3-79.

Krieg, A. (1969). Transformations in the *Bacillus cereus* - *Bacillus thuringiensis* group. Description of a new subspecies: *Bacillus thuringiensis* var. *toumanoffii*. *J. Invert. Pathol.*, 14, 279-281.

Krieg, A. (1971). Concerning exotoxin produced by vegetative cells of *Bacillus thuringiensis* and *Bacillus cereus*. *J. Invert. Pathol.*, 17, 134-135.

Krieg, A., deBarjac, H., and Bonnefoi, A. (1968). A new serotype of *Bacillus thuringiensis* isolated in Germany: *Bacillus thuringiensis* var. *darmstadiensis*. *J. Invert. Pathol.*, 10, 428-430.

Kurstak, E. (1964). Donnees sur l'epizootic bacterienne naturelle provoquee par un *Bacillus* du type *Bacillus thuringiensis* sur *Kuehniella* Zeller. *Entomophaga Mem. Ser.* No. 2, pp. 245-247.

Kushner, D. J. and Heimpel, A. M. (1957). Lecithinase production by strains of *Bacillus cereus* Fr. and Fr. pathogenic for the larch sawfly *Pristiphora erichsonii* (Htg.). *Can. J. Microbiol.*, 3, 547-551.

Labaw, L. W. (1964). The structure of *Bacillus thuringiensis* Berliner crystals. *J. Ultrastruct. Res.*, 10, 66-75.

Lacey, L. A. and Mulla, M. S. (1977). Evaluation of *Bacillus thuringiensis* as a biocide of blackfly larvae (Diptera: Simuliidae). *J. Invert. Pathol.*, 30, 46-49.

La Nove, K., Nicklas, W. J., and Williamson, J. R. (1970). Control of citric acid cycle activity in rat heart mitochondria. *J. Biol. Chem.*, 245, 102-111.

Lecadet, M. M. and Martouret, D. (1962). La toxine figuree de *B. thuringiensis*. Production enzymatique de substances solubles toxiques par injection. *C. R. Acad. Sci. Paris*, 254, 2457-2460.

Lecadet, M. M. and Dedonder, R. (1967). Enzymatic hydrolysis of the crystals of *Bacillus thuringiensis* by the proteases of *Pieris brassicae*. 1. Preparation and fractionation of the lysates. *J. Invert. Pathol.*, 9, 310-321.

Lecadet, M. M. and Martouret, D. (1967). Enzymatic hydrolysis of the crystals of *Bacillus thuringiensis* by the proteases of *Pieris brassicae*. II. Toxicity of the different fractions of the hydrolysate for larvae of *Pieris brassicae*. *J. Invert. Pathol.*, 9, 322-330.

Lecadet, M. M. and Dedonder, R. (1971). Biogenesis of the crystalline inclusion of *Bacillus thuringiensis* during sporulation. *Eur. J. Biochem.*, 23, 282-294.

Lecadet, M. M., Chevrier, G., and Dedonder, R. (1972). Analysis of a protein fraction in the spore coats of *Bacillus thuringiensis*. *Eur. J. Biochem.*, 25, 349-358.

Louloudes, S. J. and Heimpel, A. M. (1969). Mode of action of *Bacillus thuringiensis* toxic crystals in larvae of silkworm, *Bombyx mori*. *J. Invert. Pathol.*, 14, 375-380.

Lüthy, P. (1973). Self destruction of the gut epithelium: a possible expanation of the mode of action of the endotoxin of *B. thuringiensis*. *J. Invert. Pathol.*, 22, 139-140.

Lüthy, P. (1980). Insecticidal toxins of *Bacillus thuringiensis*. *FEMS Microbiology Letters*. 8, 1-7.

Martouret, D. (1961). Les toxines de *Bacillus thuringiensis* et leur processus d-action chez les larves de lepidopteres. *Symp. Phytopharm. Phytiat.* Belgium. 8, 1-14.

Mattes, O. (1927). Parasitore Krankheiten der Mehlmotten larven und Versuche uber ihre Verwendbarkeit als biologisches Bekampyungsmittel. *Sitzunberg Ges. Befoerd Gesamten Naturwiss Marburg*. 62, 381-417.

McConnell, E. and Richards, A. G. (1959). The production of *Bacillus thuringiensis* Berliner of a heat-stable substance toxic for insects. *Can. J. Microbiol.*, 5, 161-168.

McEwen, F. L., Glass, E. H., Davis, A. S., and Spittstoesser, C. M. (1960). Field tests with *Bacillus thuringiensis* for control of four lepidopterous pests. *J. Insect Physiol.*, 2, 152-164.

McMurray, W. C. and Begg, R. W. (1969). Effect of valinomycin on oxidative phosphorylation. *Arch. Biochem. Biphys.*, 84, 546-550.

Metalnikov, S. (1930). Utilisation des microbes dans la lutte contre *Lymantria* et autres insectes nuisibles. *C. R. Soc. Biol.*, 105, 535-537.

Metalnikov, S. and Toumanoff, C. (1928). Recherches experimentales sur l'infection de *Pyrausta nubilalis* par des champignons entomophytes. *C. R. Soc. Biol.*, 98, 583-584.

Metalnikov, S. and Chorine, V. (1929a). Experiments on the use of bacteria to destroy the corn borer. *Sci. Rep. Int. Corn Borer Invest*, 2, 54-59.

Metalnikov, S. and Chorine, V. (1929b). On the infection of the gypsy moth and certain other insects with *Bacterium thuringiensis*. A preliminary report. *Sci. Rep. Int. Corn Borer Invest.*, 2, 60-61.

Metalnikov, S., Hergula, B., and Strail, D. M. (1930). Experiments on the application of bacteria against the corn borer. *Sci. Rep. Int. Corn Borer Invest.*, 3, 148-151.

Miteva, V. I. (1978). Isolation of plasmid DNA from various strains of *Bacillus thuringiensis* and *Bacillus cereus*. *C. R. Acad. Sci. Bulgaria.* 31, 913-916.

Murphy, D. W., Sohi, S. S., and Fast. P. G. (1976). *Bacillus thuringiensis* enzyme-digested delta endotoxin: Effect on cultured insect cells. *Science.* 194, 954-956.

Nishiitsutsuji-Uwo, J., Ohsawa, A., and Nishimura, M. S. (1977). Factors affecting the insecticidal activity of δ-endotoxin of *Bacillus thuringiensis*. *J. Invertebr. Pathol.*, 29, 162-169.

Nishiitsutsuji-Uwo, J., Endo, Y., and Himeno, M. (1979). Mode of action of *Bacillus thuringiensis* δ-endotoxin: Effect on TN-368 cells. *J. Invertebr. Pathol.*, 34, 267-275.

Pendleton, I. R. (1968). Toxic subunits of the crystal of *Bacillus thuringiensis*. *J. Appl. Bacteriol.*, 31, 208-214.

Pendleton, I. (1970). Sodium and potassium fluxes in *Philosamia ricini* during *Bacillus thuringiensis* protein crystal intoxication. *J. Invertebr. Pathol.*, 16, 313-314.

Pendleton, I. R. (1973). Characterization of crystal protoxin and an activated toxin from *Bacillus thuringiensis* var. *entomocidus*. *J. Invertebr. Pathol.*, 21, 46-52.

Pendleton, I. R., Bernheimer, A. W., and Grushoff, P. (1973). Purification and partial characterization of hemolysins from *Bacillus thuringiensis*. *J. Invertebr. Pathol.*, 21, 131-135.

Prasad, S. S. S. V. and Shethna, Y. I. (1974). Purification, crystallization and partial characterization of the antitumour and insecticidal protein subunit from the δ-endotoxin of *Bacillus thuringiensis* var. *thuringiensis*. *Biochim. Biophys. Acta.*, 363, 558-566.

Ramakrishnan, N. (1968). Observations of the toxicity of *Bacillus thuringiensis* for the silkworm, *Bombyx mori*. *J. Invertebr. Pathol.*, 10, 449-450.

Reeves, E. L. and Garcia, C. (1971a). Pathogenicity of bicrystalliferous *Bacillus* isolate for *Aedes aegypti* and other aedine mosquito larvae. *In* "Proceedings IV International Colloquium on Insect Pathology," College Park, MD, August 1970. pp. 219-228.

Reeves, E. L. and Garcia, C. (1971b). Susceptibility of *Aedes* mosquito larvae to certain crystalliferous *Bacillus* pathogens. *Proc. Calif. Mosquito Control Assn.*, 39, 118-120.

Ribier, J. (1971). L'inclusion parasporale du *Bacillus thuringiensis* Var. *berliner* 1715: moment et site de son initiation, rapports avec l'ADN sporangial. *C. R. Acad. Sci. Paris.* 273, 1444-1447.

Ribier, J. and Lecadet, M. M. (1973). Etude ultrastructurale et cinetique de la sporulation de *Bacillus thuringiensis* var. *berliner* 1715. Remarques sur la formation de l'inclusion parasporale. *Ann. Microbiol. Inst. Pasteur.*, 123A, 311-344.

Roehrich, R. (1964). Essais de laboratoire de preparations de *Bacillus thuringiensis* Berl. contre les chenilles du Carpocapse *Laspeyresia pomonella* L. *Entomophage Colloq. Intern. Pathol. Insectes, Paris, 1962 Mem. Hors. Ser.* No. 2, p. 309.

Rogoff, M. H., Ignoffo, C., Singer, S., Gard, I., and Prieto, A. (1969). Insecticidal activity of thirty-one strains of *Bacillus* against five insect species. *J. Invertebr. Pathol.*, 14, 122-129.

Saleh, S. M., Harris, R. F., and Allen, O. N. (1970a). Recovery of *Bacillus thuringiensis* var. *thuringiensis* from field soils. *J. Invertebr. Pathol.*, 15, 55-59.

Saleh, S. M., Harris, R. F., and Allen, O. N. (1970b). Method for determining *Bacillus thuringiensis* var. *thuringiensis* Berliner in soil. *Can. J. Microbiol.*, 16, 677-680.

Sayles, W. B., Aronson, J. N., and Rosenthal, A. (1970). Small polypeptide components of the *Bacillus thuringiensis* parasporal crystalline inclusion. *Biochem. Biophys. Res. Commun.*, 41, 1126-1133.

Scherrer, P., Lüthy, P., and Trumpi, B. (1973). Production of δ-endotoxin by *Bacillus thuringiensis* as a function of glucose concentrations. *Appl. Microbiol.*, 25, 644-646.

Scherrer, P., Pillinger, R. R., and Somerville, H. J. (1974). Toxicity to Lepidoptera in the exosporium membrane spores of *Bacillus thuringiensis*. *Proc. Soc. Gen. Microbiol.*, 1, 145.

Sebesta, K. and Horska, K. (1968). Inhibition of DNA-dependent RNA polymerase by the exotoxin of *Bacillus thuringiensis* var. *gelechiae*. *Biochem. Biphys. Acta.*, 169, 281-282.

Sebesta, K., Horska, K., and Anova, J. (1969a). Isolation and properties of the insecticidal exotoxin of *Bacillus thuringiensis* var. *gelechia*. *Collect. Czech. Chem. Commun.* 34, 891-900.

Sebesta, K., Horska, K., and Ankova, J. (1969b). Inhibition of the nova RNA synthesis by the insecticidal exotoxin of *Bacillus thuringiensis* var. *gelechia*. *Collect. Czech. Chem. Commun.*, 34, 1786-1791.

Sharpe, E. S. (1976). Toxicity of the parasporal crystal of *Bacillus thuringiensis* to Japanese beetle larvae. *J. Invertebr. Pathol.*, 27, 421-422.

Shiefh, T. R., Anderson, R. F., and Rogoff, M. H. (1968). Regulation of exotoxin production in *Bacillus thuringiensis*. *Bacteriol. Proc.*, p. 6.

Smirnoff, W. A. (1968). Effects of volatile substances released by foliage of various plants on the entompathogenic *Bacillus cereus* group. *J. Invertebr. Pathol.*, 11, 513-515.

Smirnoff, W. A. (1973). Results of tests with *Bacillus thuringiensis* and chitinase on larvae of the spruce budworm. *J. Invertebr. Pathol.*, 21, 116-118.

Smirnoff, W. A. and Berlinquet, L. (1966). A substance in some commerical preparations of *Bacillus thuringiensis* var. *thuringiensis* toxic to sawfly larvae. *J. Invertebr. Pathol.*, 8, 376-381.

Smirnoff, W. A. and Valero, J. (1977). Determination of the chitinolytic activity of nine subspecies of *Bacillus thuringiensis*. *J. Invertebr. Pathol.*, 30, 265-266.

Somerville, H. J. (1973). Microbial toxins. *Ann. N. Y. Acad. Sci.*, 217, 93-108.

Somerville, H. H. (1977). The insecticidal endotoxin of *Bacillus thuringiensis*. *In* "Ponif. Acad. Sci. Scripta Varia", G. B. Marini-Bettolo (ed.). 41, 253-274.

Somerville, H. J. and Pockett, H. V. (1975). An insect toxin from spores of *Bacillus thuringiensis*. *J. Gen. Microbiol.*, 87, 359-369.

Somerville, H. J. and Swain, H. M. (1975). Temperature-dependent inhibition of polyphenylalanine formation by the exotoxin of *Bacillus thuringiensis*. *FEBS Lett.*, 54, 330.

Somerville, H. J. and Pockett, H. V. (1976). Phospholipase activity in gut juice of lepidopterous larvae. *Insect Biochem.*, 6, 351-369.

Spencer, E. Y. (1968). Comparative amino acid composition of the parasporal inclusions of five entomogenous bacteria. *J. Invertebr. Pathol.*, 444-445.

Stahly, D. P., Dingman, D. W., Irgens, R. L., Field, C. C. Feiss, M. G., and Smith, G. L. (1978a). Multiple extra-chromosomal deoxyribonucleic acid molecules in *Bacillus thuringiensis*. *FEMS Microbiol. Lett.*, 3, 139.

Stahly, D. P., Dingman, D. W., Bulla, L. A., and Aronson, A. I. (1978b). Possible origin and function of the parasporal crystals in *Bacillus thuringiensis*. *Biochem. Biophys. Res. Commun.*, 84, 581-588.

Steinhaus, E. A. (1951). Possible use of *Bacillus thuringiensis* Berliner as an aid in the biological control of the alfalfa caterpillar. *Hilgardia.* 20, 359-381.

Steinhaus, E. A. (1954). Further observations on *Bacillus thuringiensis* Berliner and other spore-forming bacteria. *Hilgardia.* 23, 1-21.

Sutter, G. R. and Raun, E. S. (1967). Histopathology of European corn borer larvae treated with *Bacillus thuringiensis*. *J. Invertebr. Pathol.*, 9, 90-103.

Talalaev, E. V. (1958). Induction of epizootic septicemia in the caterpillars of Siberian silkworm moth, *Dendrolimus sibericus* Tschtv. (Lepidoptera, Lasiocampidae). *Entomol. Rev.* USSR. 37, 557-567.

Talalaev, E. V. (1959). Bacteriological method for control of Siberian silkworm. *Trans 1st Int. Conf. Insect Pathol. Biol. Control Praka* 1958. pp. 51-57.

Talalayeva, G. B. (1967). A case of bacterial epizooty in a larval population of *Selenephera lunigera* Esp. (Lepidoptera, Lasiocampidae). *Entomol. Rev.*, 46, 190-192.

Toteson, D. C., Cook, P., Andreoli, T., and Tieffenberg, M. (1967). The effect of valinomycin on potassium and sodium permeability of HK and LK red cells. *J. Gen. Physiol.*, 50, 2513-2525.

Toumanoff, C. (1953). Description de quelques souches entomophytes de *Bacillus cereus* Frank. and Frank. avec remarques sur leur action et celle d'autres bacilles sur le jaune d'oeuf. *Ann. Inst. Pasteur.*, 85, 90-99.

Toumanoff, C. (1954). L'action de *Bacillus cereus* var. *alesti* Toum. et Vago sur les chenilles de *Galleria melonella* L. and *Hypononeuta cognatella*. *Ann. Inst. Pasteur.*, 86, 570-597.

Toumanoff, C. and LeCoroller, Y. (1959). Contributions a l'etude de *Bacillus cereus* Frank. et Frank. crystallophores et pathogenes pour les larves de Lipidopteres. *Ann. Inst. Pasteur.*, 96, 680-688.

Travers, R. S., Faust, R. M., and Reichelderfer, C. F. (1976a). Inhibition of adenosine triphosphate production by *Bacillus thuringiensis* var. *kurstaki* δ-endotoxin. *In* "Proceedings of the First International Colloquium on Invertebrate Pathology," Queen's University, Kingston, Canada, pp. 408-409.

Travers, R. S., Faust, R. M., and Reichelderfer, C. F. (1976b). Effects of *Bacillus thuringiensis* var. *Kurstaki* δ-endotoxin on isolated lepidopteran mitochondria. *J. Invertebr. Pathol.*, 28, 351-356.

Treherne, J. E. (1958a). The absorption of glucose from the alimentary canal of the locust, *Schistocerca gregaria* Forsk. *J. Exp. Biol.*, 35, 297-306.

Treherne, J. E. (1958b). The absorption and metabolism of some sugars in the locust, *Schistocerca gregaria* Forsk. *J. Exp. Biol.*, 35, 611-625.

Van der Laan, P. A. and Wassink, H. J. M. (1964). Susceptibility of different species of stored-products moth larvae to *Bacillus thuringiensis*. *Colloquium International Patholgoy Insectes, Paris, 1962*. *Entomophaga, Memo.*, 2, 315-322.

Vankova, J. (1964). *Bacillus thuringiensis* in praktischer Anwendung. *Entomophaga.* 2, 271-291.

Yamvrias, C. (1961). Contribution a ētude der mode l'action de *Bacillius thuringiensis* Berliner vis-a-vis de la teigne de la farine: *Anagasta (Ephestia) kühniella* Zeller. *Doctoral thesis*, Universite of Paris, Paris.

Young, I. E. and Fitz-James, P. C. (1959). Chemical and morphological studies of bacterial spore formation. II. Spore and parasporal protein formation in *Bacillus cereus* var. *alesti*. *J. Biophys. Biochem. Cytol.*, 6, 483-498.

Yousten, A. A. and Guerran, R. L. (1976). Comparative activities of *Bacillus thuringiensis* δ-endotoxin, cholera toxin, and *Escherichia coli* enterotoxin. *J. Invertebr. Pathol.*, 28, 395-397.

BIOLOGICAL CONTROL OF THE PSYCHODIDAE, CERATOPOGONIDAE, AND SIMULIIDAE (DIPTERA), INCLUDING NEW DATA ON THE EFFECT OF *Bacillus thuringiensis* VAR. *israelensis* ON SIMULIIDAE

Albert H. Undeen

Insects Affecting Man and Animals Research Laboratory
Agricultural Research Service
U.S. Department of Agriculture
Gainesville, Florida

I. INTRODUCTION

Biological control of the Psychodidae (moth flies and sand flies) and the Ceratopogonidae (biting midges) by pathogens or parasites will require a more intensive research effort in the future if their potential as biocontrol agents is to be realized. Numerous pathogenic and parasitic organisms have been described associated with these Diptera, but limited effort has been made to study and develop them for use. Indeed, an active search for biocontrol agents of these two families must be a continuing effort. In contrast, a number of pathogens and parasites affecting the Simuliidae (blackflies) have been described and tested successfully, especially *Bacillus thuringiensis* var. *israelensis* (see Undeen and Nagel, 1978). Other potential agents for biocontrol of the Simuliidae include viruses, fungi, microsporidia, and mermithid nematodes.

In general three major factors should be considered in the development of biocontrol agents for pest insect control: (1) the agent must be infectious, (2) the agent must be deleterious to the host population, and (3) the agent must be amenable to mass production.

II. PATHOGENS AND PARASITES OF PSYCHODIDAE

Most pathogens and parasites of psychodids have been isolated from the adult insect because it is the stage most accessible for examination. Table 1 lists the known potential pathogens and parasites of the Psychodidae. Obviously, an intensive search for pathogens and parasites affecting the larval stage is greatly needed. At the present time, the *Entomophthora* fungi are the only probable candidates for biocontrol of psychodids and apparently they affect mainly the adult stage (Roberts and Strand, 1977). None of the agents listed, other than *Aspergillus* and *Penicillium*, have been experimentally transmitted to the natural host or reared under laboratory conditions. In fact, even the pathogens that infect the Ceratopogonidae, such as *Coelomycidium* and most of the Microspordia affecting blackflies and mosquitoes, have never been successfully retransmitted to their original hosts (Goldberg, 1971). Most of the agents in Table 1 also have not been shown to meet the three major requirements necessary to be effective biological control agents. Efforts similar to work described for mosquitoes and blackflies, especially with their mermithids, should be carried out with the Psychodidae.

III. PATHOGENS AND PARASITES OF CERATOPOGONIDAE

One of the most complete reports on pathogens and parasites of the ceratopogonids has been prepared by Yaseen (1971). Unfortunately, the report is weighted heavily towards the incidence of parasitemia. One exception that can be noted is the description of the fungus *Grubyella ochoterenai* that infects blackflies. It could be experimentally transmitted to midge larvae under laboratory conditions resulting in the death of the flies. Table 2 lists the known pathogens and parasites of the Ceratopogonidae.

Since ceratopogonid larvae have diverse feeding habits, special considerations in the selection of potential biocontrol agents are required. For example, some ceratopogonid species are predominately predacious and may not be highly susceptible

TABLE 1. Known potential pathogens and parasites affecting the Psychodidae.[a]

PATHOGEN	STAGE INFECTED	DEMONSTRATED PATHOGENICITY	CULTURED IN THE LABORATORY	EXPERIMENTALLY TRANSMITTED
Rickettsia sp	adult	-	-	-
Spirochaeta phlebotomi	adult	-	-	-
Pseudomonas sp	larva and pupa	+	-	-
Aspergillus	larva	+	+	+
Penicillium glaucum	larva	+	+	+
Coelomomyces citerrii	egg	+	-	-
Entomophthora sp	adult	+	-	-
Fungus	adult	+	-	-
Gregarinida	larva pupa and adult	-	-	-
Microsporidia	adult	-	-	-
Mermithids	adult	+	-	-

[a] Compiled from Roberts and Strand (1977).

TABLE 2. Known potential pathogens and parasites affecting the Ceratopogonidae.

PATHOGEN	STAGE INFECTED	DEMONSTRATED PATHOGENICITY	CULTURED IN THE LABORATORY	EXPERIMENTALLY TRANSMITTED
Iridescent virus[a]	larva	+	-	-
Rickettsia sp.[a]	adult	-	-	-
Bacteria[a]	larva and adult	-	-	-
Psendomonas sp.[a]	larva	-	-	-
Monosporella unicuspidata[a]	larva	+	-	-
Grubyella ochoterenai[a]	adult	+	-	+
Entomophthora[a]	adult	+	+	-
Gregarinida[b]	larva	-	-	-
Microsporidia[b]	larva	+	-	-
Ciliata[b]	adult	-	-	-
Helicosporidium parasiticum[a]	larva	+	-	-
Mermithids[b,c]	larva and adult	+	-	-

[a]Compiled from Roberts and Strand (1977).
[b]Rebholz et al., 1977.
[c]Phelps and Mokry, 1976.

to enteric bacteria as would a detritus feeder. Furthermore, the larvae of thse midges develop in a variety of aquatic habitats, especially the salt marsh. Salt marsh environments are not very amenable to chemical pesticide application because of their close ecological relationship to the habitat of coastal fishes and shellfish. The development of agents for biocontrol of ceratopogonids in these sensitive areas is highly desirable and less likely to disrupt its ecological base.

IV. PATHOGENS AND PARASITES OF SIMULIIDAE

The identification of pathogens and parasites affecting the Simuliidae has been relatively extensive as indicated by Table 3. The range of potential biocontrol agents include bacteria, viruses, fungi, protozoans, and nematodes.

The nematode *Romanomermis culcivorax* a mosquito parasite, has been reported capable of invading and killing blackfly larvae but high concentrations of preparasites are necessary, even in still water (Finney, 1975). Molloy and Jamnback (1977) released preparasites of *Neomesomermis flumenalis* into a small stream harboring *Simulium venustum* and *S. vittatum* larvae and found an incidence of infection as high as 71.4% in *S. vittatum*. *N. flumenalis* also was found to infect two other species of *Simulium* in this study. The preparasites had been field collected from mature worms of *Prosimulium* sp. that had been held in the laboratory during post-parasitic development, oviposition, and hatching (Molloy and Jamnback, 1975). This method for obtaining nematodes, of course, is neither a practical nor an economical means for controlling blackfly larvae, but it does indicate that these agents could be make useful provided an economical means for their mass production was developed. With our present technology, *in vivo* mass rearing of simuliid mermithids is technically possible, but not economical at this time. Considerable more basic research is necessary before an *in vitro* production method is possible.

The habitat of blackfly larvae poses several problems for the use of biocontrol agents. For example, the rapidly-flowing water would disperse and transport the applied material downstream in a relatively short time period, thus making it unavailable to the target insect. Fortunately, mermithid worms and the ovarian phycomycete are both capable of upstream dispersal in the adult flies. The microsporidia are probably transmitted through the egg, though transovarian transmission has not been definitively proven. The complete life cycle of the microsporidian *Coelomycidium*, the ciliates, and the ovarian phycomycete need to be further worked out.

TABLE 3. Known potential pathogens and parasites affecting the Simuliidae.

PATHOGEN	STAGE INFECTED	DEMONSTRATED PATHOGENICITY	CULTURED IN THE LABORATORY	EXPERIMENTALLY TRANSMITTED
Cytoplasmic polyhedrosis virus[a]	larva and adult	+	–	+
Iridescent virus[b]	larva	–	–	–
Parovirus[b]	larva	+	–	–
Densonucleosis virus[c]	larva	–	–	–
Bacterium (unidentified)	larva and adult	+	+	–
Bacillus sphaericus[b]	larva	–	+	+
Bacillus thuringiensis (Lepidoptera)[e]	larva	+	+	+
Bacillus thuringiensis var. *israelensis*[f]	larva	+	+	+
Trichomycetes[b]	larva	–	+	+
Coelomycidium simulii[g]	larva	+	–	–
Saprolengniacea[b]	larva	+	+	+
Aspergillus sp.[b]	larva	+	+	+
Entomophthora[b]	adult	+	+	+
Ovarian phycomycete[h]	adult	+	–	–
Microsporidia[b]	larva	+	–	–

PATHOGEN	STAGE INFECTED	DEMONSTRATED PATHOGENICITY	CULTURED IN THE LABORATORY	EXPERIMENTALLY TRANSMITTED
Ciliata[b]	adult	-	-	-
Gregarinida[b]	adult	+	-	-
Haplosporidium simulii[b]	larva	+	-	-
Mermithids[b]	larva and adult	+	+	+

[a] Bailey, 1977.
[b] Compiled from Roberts and Strand (1977).
[c] Frederici and Lacey, 1976.
[d] RUVP Animal Reports, 1977, 1978.
[e] Lacey and Mulla, 1977.
[f] Undeen and Nagel, 1978.
[g] Levchenko and Dzerzhinskii, 1976.
[h] Undeen and Nolan, 1977.

Experimental transmission of a cytoplasmic polyhedrosis virus (CPV) has been reported but mortality was low, though surviving adults transmitted the virus to the next generation of blackflies (Bailey, 1977). *Cnephia mutata* larvae collected in the field apparently succumb in the laboratory from latent CPV infection. Mass production of the CPV affecting blackflies has yet to be accomplished.

Certain fungi (e.g., *Saprolegniaceai* and *Aspergillus*) have been shown to be pathogenic to blackfly larvae. However, they exhibit a broad spectrum of activity and may have difficulty being registered as a biorational pesticide. The Trichomycetes occasionally infect 100% of blackfly larvae in a population but fail to kill the host. The infectious stage of the ovarian phycomycete has not been isolated. However, natural incidences in excess of 60% infection have been described for some ovipositing populations of *C. mutata*; none of the infected flies produced eggs - at least in the first cycle (Undeen and Nolan, 1977).

The Research Unit on Vector Pathology, Memorial University of Newfoundland, is currently testing blackfly entomopathogens which are easily obtainable and amenable to laboratory cultivation (pers. commun.). Several microsporidia species (e.g., *Nosema algerae*, *N. eurytraemae*, and *Vairimorpha* [= *Nosema*] *necatrix*) tested have failed to infect the larvae although the spores germinate in the gut. *B. thuringiensis*, a pathogen mainly of lepidopteran larvae, has been reported to be pathogenic for *S. vittatum* (Lacey and Mulla, 1977) and the Research Unit on Vector Pathology has screened strains of *B. thuringiensis* with similar results. The Research Unit on Vector Pathology has also tested a strain of *B. sphaericus* obtained from Dr. Samuel Singer, Western Illinois University, Macolm, Illinois. Blackflies were not susceptible to this strain. Purified spore-crystal preparations of a *B. thuringiensis* strain obtained from Dr. Paul Fast, Sault St. Marie, Canada, caused no mortality at doses above 10^7 spores/ml. However, a relatively new strain of *B. thuringiensis* (var. *israelensis*) (deBarjac, 1978) was found to be highly effective against blackfly larvae. Table 4 summarizes the results of 30-min exposures to larvae of several Newfoundland blackfly species to *B. thuringiensis* var. *israelensis* (Undeen and Nagel, 1978). Efficacy tests in small streams later verified our laboratory results (Undeen, unpubl.). Lüthy et al. (1980) have recently reported the field efficacy of *B. thuringiensis* var. *israelensis* in the Swiss canton of Wallis for control of three species of *Aedes*. The main experiment, conducted in a wildlife reservation, resulted in a total mortality of the mosquito larvae within 15 days. This bacterial pathogen also should be tested against the Psychodidae and the Ceratopogonidae.

TABLE 4. Effect of 30-minute exposure of various blackfly larvae species to *Bacillus thuringiensis* var. *israelensis*.[a,b]

SPECIES	LC_{50}	LC_{90}
Cnephia ornithophilia	1.5×10^3	2.5×10^3
Simulium verecundum	2.5×10^3	5.6×10^3
Mixture of:		
Cnephia mutata *Cnephia ornithophilia* *Simulium vittatum* *Prosimulium mixtum*	1.8×10^3	5.0×10^3

[a] LC_{50} and LC_{90} in viable spores/ml.

[b] Undeen and Nagel, 1978.

V. CONCLUSIONS

Research on the biological control of the nematoceran Diptera, other than mosquitoes, is at an early developmental stage. More intensive research is necessary before any pathogen or parasite can be seriously considered for the biocontrol of Psychodidae or Ceratopogonidae and, in most cases, the Simuliidae. A more thorough search must be initiated for microbial pathogens of these flies. Infectivity and pathogenicity studies are needed on the few agents that have been described. Screening for effectiveness of existing microbes from other hosts could be advantageous because laboratory cultivation and safety data are already available.

Biocontrol efforts for blackflies is somewhat more advanced. Research has been described for the use of *Neomesomermis flumenalis*, a mermithid nematode. In one small-scale field test, incidence of infection reached 70% (Molloy and Jamback, 1977). *Bacillus thuringiensis* var. *israelensis*, a candidate for microbial control of mosquitoes is pathogenic to simuliid larvae with LC_{90}s of 5×10^3 viable spores/ml. Field application of *B. thuringiensis* var. *israelensis* spores produced mortality comparable with laboratory data.

In the search for an organism to use for biocontrol of biting flies it might be profitable to test all available entomopathogens against each pest group. Pathogen specificities are often not absolute and efficiency can be promoted if an agent that already has a background of biocontrol testing can be found to have a new application.

VI. REFERENCES

Bailey, C. H. (1977). Field and laboratory observations on a cytoplasmic polyhedrosis virus of blackflies (Diptera: Simuliidae). *J. Invertebr. Pathol.*, 29, 69-73.

deBarjac, H. (1978). Une nouvelle variété de *Bacillus thuringiensis* trés toxique pour les moustiques: *B. thuringiensis* var. *israelensis* sérotype 14. *C. R. Acad. Sc. Paris*, 28, 797-800.

Federici, B. A. and Lacey, L. A. (1976). Densonucleosis virus and cytoplasmic polyhedrosis virus diseases of the blackfly, *Simulium vittatum*. *Proc. Pap. Annu. Conf. Calif. Mosq. Control Assoc.*, 44, 124.

Finney, J. R. (1975). The penetration of three simuliid species by the nematode *Reesimermis nielseni*. *Bull. Wld. Hlth. Org.*, 52, 235.

Goldberg, A. M. (1971). Methods of isolating and cultivating fungi of the family Entomophthoraceae, Phycomycetes, parasitizing mosquitoes of the family Culicidae and biting midges of the family Ceratopogonidae. *Minerva Fisioterapica*, 5, 518-522.

Lacey, L. A. and Mulla, M. S. (1977). Evaluation of *Bacillus thuringiensis* as a biocide of blackfly larvae (Diptera: Simuliidae). *J. Invertebr. Pathol.*, 30, 46-49.

Levchenko, N. G. and Dzerzhinskii, V. A. (1976). Harmfulness of the aquatic fungus *Coelomycidium simulii* (new record) parasitizing blackfly larvae (*Odagmia* sp.). *Izv. Akad. Nauk Kaz. SSR. Ser. Biol.*, 14, 18-21.

Lüthy, P., Raboud, G., Delucchi, V., and Kuenzi, M. (1980). Field efficacy of *Bacillus thuringiensis* var. *israelensis*. *Bull. Soc. Entomol. Suisse*, 53, 3-9.

Molloy, D. and Jamnback, H. (1975). Laboratory transmission of mermithids parasitic in blackflies. *Mosq. News*, 35, 337-342.

Molloy, D. and Jamnback, H. (1977). A larval black fly control field trial using mermithid parasites and its cost implications. *Mosq. News*, 37, 104-108.

Phelps, R. J. and Mokry, J. E. (1976). *Culicoides* species (Diptera: Ceratopogonidae) as potential vectors of filariasis in Rhodesia. *J. Ent. Soc. South Afr.*, 39, 201-206.

Rebholtz, C., Zenner, F., and Kremer, M. (1977). Concerning some parasites found in *Culicoides*. *Mosq. News*, 37, 284.

Roberts, D. W. and Strand, M. A. (1977). Pathogens of medically important arthropods. *Bull. Wld. Hlth. Org.*, 55, (Suppl. No. 1), 419 pp.

RUVP Annual Reports (1972-1978). Memorial University of Newfoundland, St. John's, Newfoundland, Canada.

Undeen, A. H. and Nolan, R. A. (1977). Ovarian infection and fungal spore oviposition in the blackfly, *Prosimulium mixtum*. *J. Invertebr. Pathol.*, 30, 97-98.

Undeen, A. H. and Nagel, W. L. (1978). The effect of *Bacillus thuringiensis* ONR - 60A strain (Goldberg) on *Simulium* larvae in the laboratory. *Mosq. News*, 38, 524-527.

Yaseen, M. (1971). Investigations into the possibilities of biological control of sandflies (Diptera: Ceratopogonidae). *Rep. Commonw. Inst. Biol. Contr.*, (Final Report) p. 1-14.

THE *Bacillus thuringiensis* GENETIC SYSTEMS

Burton D. Clark, Frederick J. Perlak, Cho-Yam Chu,
and Donald H. Dean

Dept. of Microbiology, Denetics and The Program in
Molecular, Cellular and Developmental Biology
The Ohio State University
Columbus, Ohio

I. INTRODUCTION

The last few years have seen a major effort to improve the genetics of *Bacillus thuringiensis*, reminiscent of the tenacity with which Edward A. Steinhaus pursued his early research on this insect pathogen (Steinhaus, 1975). This review is in memory of his dedication to the study of biological control of pests.

The driving question in genetic research on *B. thuringiensis* has been the location of the crystal toxin (δ-endotoxin) gene. Of course, this question would have no meaning had not plasmids been observed in several varieties (Zakharyan et al., 1976; Debabov et al., 1977; Ermakova et al., 1978; Stahly et al., 1978a; Faust et al., 1979; Yousten, 1978; Gonzalez and Carlton, 1980; Martin and Dean, 1980; Martin et al., 1981; Iizuka et al., 1981a), prompting speculation that the crystal toxin gene is on a plasmid. Unfortunately, there has not been an easy answer to this question nor has the genetic manipulation of *B. thuringiensis* been easy. But these discouragements have only drawn the geneticist deeper into the challenge of this intriguing bacterium. This paper will critically review the genetic systems available for the study of *B. thuringiensis* and address the possibility that the crystal toxin gene is plasmid-borne.

II. GENERALIZED TRANSDUCTION

CP-51 and CP-54 Transducing Phages. Several bacteriophages have been shown to be capable of generalized transduction in *B. thuringiensis* (Thorne, 1978; Perlak et al., 1979). CP-51, a generalized transducing phage of *B. cereus* (Thorne, 1968a,b), was reported to mediate transduction between *B. thuringiensis* 1328 (var. unknown) as the donor and *B. cereus* 6464 and *B. cereus* 569 as recipients (Yelton and Thorne, 1970). Thorne (1978) utilized CP-51 and another phage isolated from soil (CP-54) to mediate generalized transduction among *B. thuringiensis* strains. CP-51 was lytic for all varieties of *B. thuringiensis* except var. *finitimus*, var. *alesti*, var. *galleriae*, and var. *entomocidus-limassol*, whereas CP-54 lysed all *B. thuringiensis* strains tested. However, CP-51 and CP-54 were too lytic on some strains to be useful as transducing phages, even after inactivation of plaque forming ability with ultraviolet light. Interstrain transduction was successful with CP-54 only with var. *finitimus*, var. *alesti*, var. *sotto*, var. *galleriae*, var. *canadensis*, and var. *entomocidus-limassol* (donors), and var. *finitimus* or *alesti* (recipients). Recently, Lecadet (1980) isolated a variant of CP-54 (called CP-

54-Ber) which mediates generalized transduction in var. *thuringiensis* Berliner 1715.

TP-10, TP-11, TP-13, and TP-18 Tranducing Phages. Other bacteriophages which are capable of generalized transduction in *B. thuringiensis* are TP-10, TP-11, TP-12, TP-13, and TP-18 (C. B. Thorne, pers. commun.). These bacteriophages have not been fully characterized.

Bacteriophage TP-13 was isolated from soil and has many characteristics of a good generalized transducing phage (Perlak et al., 1979). It is a phage with a wide host range among the *B. thuringiensis* varieties and it appears to be analogous to the large generalized transducing phages (PBS-1, SP-15) of *B. subtilis*. These three phages are similar in their size and morphology, capability of generalized transduction, and requirement for motile cells for infection.

The large size of TP-13 suggests its utility for chromosomal mapping. Transduction frequencies of *B. thuringiensis* with TP-13 range from 10^{-6} to 10^{-5} per plaque forming unit (PFU). Comparison of the co-transduction values of several markers with TP-10, CP-51, and TP-13 indicates that the DNA fragments transduced by the three phages are different in size (Perlak et al., 1979). Together they comprise an attractive genetic analysis system useful for scanning for chromosome, as well as for fine structure mapping. Table 1 is a list of the reported generalized transducing phages of *B. thuringiensis*.

TP-13 has been shown to convert an oligosporogenic, acrystalliferous mutant of *B. thuringiensis* (UM17, *spo-1*) to crystal^{+} spore^{+} at a high frequency (Perlak et al., 1979). The simultaneous conversion to sporulation and crystal formation by TP-13 indicates a close relationship between the two processes. Previous studies have implied a relationship between sporulation and crystal formation (Delafield et al., 1968). A common glycoprotein subunit has been found in the crystal protein and the spore coat (Bulla et al., 1977) and mutants have been isolated which are simultaneously acrystalliferous and asporogenous (Somerville, 1968). TP-13 mediated generalized transduction has linked the *spo-1* mutation to the *met-1* marker (F. Perlak, unpubl.). The apparent chromosomal location of *spo-1* suggests that some function required for the production of the crystal protein involves chromosomal genes. This finding does not establish that all genes for crystal production are on the chromosome.

Several groups of markers have been linked by transduction with TP-13 (F. Perlak, unpubl.). These linkages do not appear to be analogous to the genetic map of *B. subtilis*. Genetic mapping of auxotrophic mutations and mutations conferring

TABLE 1. List of generalized transducing phages for *Bacillus thuringiensis*.

PHAGE	REFERENCE
Thl	deBarjac, 1970
CP-51	Thorne, 1978
CP-54	Thorne, 1978
CP-54-Ber	Lecadet, 1980
TP-10	Perlak et al., 1979
TP-11	Thorne (pers. commun.)
TP-12	Thorne (pers. commun.)
TP-13	Perlak et al., 1979
TP-18	Thorne (pers. commun.)

asporogenous and acrystalliferous phenotypes in *B. thuringiensis* will hopefully give new insights into the relationship between crystal formation and sporulation.

III. TRANSFORMATION

Classical Transformation. The early discovery of transformation in *B. subtilis* (see Spizizen, 1958) gave promise that a similar genetic system might be available in other *Bacillus* species. To date, however, there is only one report of successful transformation of *B. thuringiensis* with classical procedures. Reeves (1966) used the method of Spizizen (1958, 1959, 1961) to transform an avirulent species *B. finitimus* (i.e., *B. thuringiensis* var. *finitimus*) with DNA from the virulent species *B. thuringiensis* var. *thuringiensis*. Since there was no simple selection system to determine the success of transformation, Reeves picked colonies which had been transformed to the morphology of the donor type and screened for pathogenesis to *Trichoplusia ni* (cabbage looper). He reported a five-fold increase in pathogenesis of the variety *finitimus*.

Polyethylene Glycol-Induced Protoplast (PIP) Transformation. The potential to transform cells, which are not able to take up purified DNA by natural means, was developed in experiments by Bibb et al. (1978) for the genus *Streptomyces*. The procedure involves producing cell protoplasts and mixing them with DNA in the presence of polyethlene glycol (PEF). PEG binds the DNA to the cell allowing it to enter and effect transformation. This method is particularly useful for the introduction of plasmids into an organism.

The first application of the polyethylene glycol-induced protoplast (PIP) transformation on *Bacillus (B. subtilis)* was by Chang and Cohen (1979). Ryabchenko et al. (1980) transformed *B. thuringinesis* var. *galleriae* 69-6 with the plasmid pBC16, a tetracycline resistance plasmid from *B. cereus*. This article was published in English by Alikhanian et al. (1981). Independently, Martin et al. (1981) reported introducing a chloramphenicol resistant plasmid, pC194, from *Staphylococcus aureus* into a derivative of the commercial strain of *B. thuringiensis* var. *kurstaki* by PIP transformation. The plasmid pC194 is a certified cloning vehicle for *B. subtilis* and has the capabilities of accepting recombinant DNA. Martin et al. (1981) also observed that pC194 has the unusual property of inserting into the *B. thuringiensis* chromosome. Further research indicates that pC194 inserts 4 to 8 times (Lohr and Dean, unpubl.). The discovery of ways to introduce plasmid cloning vehicles into *B. thuringiensis* opens the possibilities of using recombinant DNA to the field of biological control of insects.

IV. PLASMIDS AND CRYSTALS

Plasmid Isolation and Partial Characterization. *B. thuringiensis* is a commercially important bacterium because of parasporal crystal with entomopathogenic properties. Very little is known concerning the gene(s) encoding for the production of this crystal. The frequency of crystal production lost is higher than that of ordinary spontaneous mutants (Toumamoff, 1955; Vankova, 1957; Fitz-James and Young, 1959). Acrystalliferous strains are often isolated after heat treatment of the spores. Such phenotypic losses can occur through the loss of an unstable plasmid (Lacey, 1975). Reversion from an acrystalliferous phenotype to a crystal producing phenotype is extremely rare (Norris, 1970).

In attempts at attributing crystal production and other biochemical functions to plasmids in *B. thuringiensis*, plasmid DNA has been isolated and partially characterized. The first study suggesting that plasmids play a role in coding for the biosynthesis of the crystal was by Zakharyan et al. (1976). They later isolated three plasmids from *B. thuringiensis* var. *caucasicus* (Zakharyan et al., 1979). The plasmid DNA was purified by ultracentrifugation and separated by electrophoresis in 0.6% agarose gels. The molecular weights were determined to be 6, 10, and 90 megadaltons. The plasmids, isolated from the agarose gel, were restricted with *Eco*RI. No other studies have since reported restriction analysis of plasmids in *B. thuringiensis*. Stahly et al. (1978a) examined the plasmids in *B. thuringiensis* var. *alesti*. Twelve plasmids were examined by electron microscopy and the molecular weights determined from measurment of the contour length compared to that of ΦX 174 RF DNA. From the molecular weights reported, some of the smaller plasmids may be present in dimeric and trimeric forms, a common observation with Col E1 and pBR322 in *E. coli* and pC194 in *S. aureus* and *B. subtilis*.

Miteva (1978) reported the first extensive study to detect the presence of plasmids in serotypes 1-11. Most of the plasmids isolated in these serotypes have molecular weights ranging from 3-9 megadaltons. The data show the presence of both covalently closed circular (CCC) forms as well as open circular (OC) forms even though the plasmid DNA was isolated by CsCl-ethidium bromide gradients which is selective for the CCC form. In a recent study by Iizuka et al. (1981a), clear lysates of all 17 serotypes were examined for the presence of plasmids by agarose gel electrophoresis. The number of plasmid bands based on the agarose gel electrophoresis profiles ranged from one for *B. thuringiensis* var. *sotto* and var. *thompsoni*, to 16 for *B. thuringiensis* var. *kurstaki*. These plasmid profiles consist of both CCC and OC forms. The molecular weights of the CCC form range from less than 1 megadalton to greater than 120 megadaltons. This study included the isolation of plasmids from three new serotypes: *B. thuringiensis* var. *dakota*, var. *indiana*, and var. *wuhanensis*. It is interesting that all the strains representative of the 17 serotypes possess at least one and as many as 16 plasmids. Such diversity has not been seen in *B. subtilis* or *B. pumilus* (Lovett and Bramucci, 1975; Tanaka et al., 1977). Lovett and Bramucci (1975) found plasmids in only two out of 18 strains of *B. subtilis* screened, and in three out of 20 strains for *B. pumilus* screened. Tanaka (1977) observed similar results in *B. subtilis*; plasmids were found in only four out of 19 strains screened.

Effects of Growth Medium on Plasmids and Crystal Production. Ermakova et al. (1978) observed a difference in the quantity of plasmid DNA and crystal production in *B. thuringiensis* var. *galleriae*, depending on the composition of growth medium. When plasmid DNA is isolated from cells grown in nutrient media, three plasmids with molecular weights of 5.9, 10, and 10.9 megadaltons are seen using electron microscopy and agarose gel electrophoresis. They are similar to the values reported by Debabov et al. (1977). Cells grown in nutrient medium for 48-72 hr produced crystals. If the same strain is grown using Spizizen minimal salts supplemented with 0.2% sodium citrate (Spizizen, 1958; Nickerson and Bulla, 1974), there is a reduced ability to form crystals and the disappearance of plasmid DNA. The plasmid DNA fraction reappears on transfer from Spizizen minimal salts to nutrient media. The sensitivity of plasmid DNA detection was 0.3 μg DNA per ultracentrifugation tube after photography in ultraviolet light (Ermakova et al., 1978). A plausible explanation for the disappearance of plasmid DNA in cells grown in Spizizen minimal salts is a decrease in plasmid copy number below the sensitivity of the method of plasmid DNA detection utilized. The results suggest that there may be a correlation between the presence of plasmid DNA and the formation of crystals. The decreased ability to form crystals may therefore be due to a decrease in the copy number of a plasmid-borne crystal gene.

Comparison of Plasmid Profiles in Acrystalliferous and Crystalliferous Strains. There are several studies that have investigated the possibility of plasmid DNA encoding for parasporal crystal production by comparing the plasmid profiles of crystalliferous and acrystalliferous strains (Miteva, 1978; Stahly et al., 1978b; Gonzalez and Carlton, 1980). For example, Miteva (1978) isolated plasmid DNA from two mutants (*spo*$^+$*cry*$^-$ and *spo*$^-$*cry*$^-$) of *B. thuringiensis* var. *thuringiensis* and compared them with the parental type. There was no difference between the plasmid profiles. These observations do not exclude the possibility of a plasmid-encoded crystal. The acrystalliferous phenotype may result from a mutation within the gene(s) encoding for the crystal protein. Stahly et al. (1978b), Yousten et al. (1978), and Martin (1979) cured plasmids from *B. thuringiensis* var. *kurstaki* and concomitantly induced acrystalliferous mutants by treating spores with heat. The frequency of acrystalliferous mutants obtained by this method varied from 1 to 2 x 10^{-3}. Stahly et al. (1978b) reported that in the parental strain, six plasmids could be differentiated by contour analysis of electron micrographs and that there were no detectable plasmids in the three acrystalliferous mutants. Recent studies in our laboratory have confirmed the observation that plasmids are absent in one of these mutants (*cry-B*). We have not yet examined the plasmids in the other two mutants (S_1 and R_6). Plasmid DNA patterns in crystalliferous and acrystalliferous strains of both *B. thuringiensis* var. *kurstaki* and var. *thuringiensis* have been compared by Gonzalez and Carlton (1980). Eleven distinct plasmids ranging in molecular weight from 1.9 to 52 megadaltons were isolated from var. *kurstaki* (HD-1). The acrystalliferous strain (HD-31) had a similar plasmid profile but some plasmids were absent. In contrast, the results reported by Stahly et al. (1978b) indicate that the three acrystalliferous strains are accompanied by the loss of all plasmid DNA. It is not possible to associate a plasmid with crystal production when several plasmids are missing. The comparison of several acrystalliferous derivatives of *B. thuringiensis* var. *thuringiensis* reveal a consistent loss of a large plasmid of approximately 75 megadaltons (Gonzalez and Carlton, 1980).

Recently, a similar study compared the plasmid profile of single and multiple crystalliferous strains of *B. thuringiensis* var. *kurstaki* (see Iizuka et al., 1981b). Normally only one crystal per cell is seen in *B. thuringiensis*. Exceptions are *B. thuringiensis* var. *darmstadiensis* (see Krieg et al., 1968), *B. thuringiensis*

BA-068 (see Reeves and Garcia, 1971), and *B. thuringiensis* var. *israelensis* (see Lüthy, 1980). Recently a strain of *B. thuringiensis* var. *kurstaki* has been isolated which produces 2-5 crystals per cell (S. Amonkar, pers. commun.). Plasmid DNA from the multicrystalliferous strain was compared to *B. thuringiensis* var. *kurstaki* (HD-1 and HD-73) with hopes of an extra plasmid or an increase in copy number of a particular plasmid. An 18 megadalton plasmid is seen in the multicrystalliferous strain that is not seen in HD-1 or HD-73. However, other differences in the plasmid profile made it difficult to associate plasmid DNA with crystal production (Iizuka et al., 1980b).

Crystal Protein, Stable mRNA, and Plasmids in B. thuringiensis var. thuringiensis. Extensive studies have been conducted concerning the existance of stable and unstable mRNA and distinct enzymatic forms of RNA polymerase in *B. thuringiensis* var. *thuringiensis* (see Glatron and Rapoport, 1972, 1975; Rain-Guion et al., 1976; Klier and Lecadet, 1976; and Klier et al., 1978). Treatment of vegetative cells with Rifampin suppressed incorporation of ^{3}H-valine by 95% in less than 10 min at 30°C, but the drug did not completely inhibit incorporation of the labeled amino acid in sporulating cells. The residual protein synthesis after treatment of the cells with Rifampin was attributed to stable mRNA, some encoding for the crystal protein. Direct incorporation of ^{3}H-valine was demonstrated in the crystal protein after immunoprecipitation with specific antibodies (Glatron and Rapoport, 1972). Comparing the amount of ^{3}H-valine incorporated into the crystal protein to that incorporated into the total bacterial protein after various times of treatment with Rifampin indicated that the crystal protein represented 40% of the protein coded by the stable mRNA fraction.

It has been suggested that the increased stability of the mRNA during sporulation is due to protection acquired by close association with membrane structures and polysomes (Aronson, 1965). Electron micrographs of cells treated and cells not treated with Rifampin show crystal inclusions bound to the membrane and surrounded by ribosomes. The mRNA coding for crystal protein may be protected by this cell structure (Glatron and Rapoport, 1975; Ribier and Lecadet, 1973).

The stable mRNA coding for the crystal protein was enriched by sucrose gradient centrifugation. Fractions of stable mRNA sediment around 19S, between the 16S rRNA and 23S rRNA. Competition using different *in vivo* RNA species suggests that the 19S RNA fraction has sequences corresponding to the crystal protein mRNA. The mRNA in the 19S fraction is preferentially transcribed by the sporulation Form II RNA polymerase from the L strand of the DNA (Klier et al., 1978). Having established the association between stable mRNA and the crystal protein in *B. thuringiensis* var. *thuringiensis* Berliner

1715, Klier et al. (1978) have recently used radioactively labelled stable mRNA to probe the location of the crystal protein gene. They have found that the probe hybridizes to the chromosome and not to the plasmids of this strain.

Plasmids in B. thuringiensis var. israelensis. B. *thuringiensis* var. *israelensis* was first reported by Goldberg and Margalit (1977) as a new strain of B. *thuringiensis* which is highly toxic to mosquito larvae. The strain was further characterized by deBarjac (1978a) and designated as serotype H-14. Unlike other varieties of B. *thuringiensis* which are mainly toxic to lepidoptera larvae, var. *israelensis* exhibits high larvicidal activity against many species of Diptera (Goldberg and Margalit, 1977; deBarjac, 1978b; van Essen, and Hembree, 1980; Undeen and Colbo, 1980). The toxicity is the result of action of the parasporal crystal toxin (deBarjac and Larget, 1979).

As with other varieties of B. *thuringiensis*, the genetics of var. *israelensis* is in its infancy. The plasmid profile of var. *israelensis* was recently reported by Iizuka et al. (1981a). Using a SDS-high salt method to prepare the cleared lysate, they were able to identify six plasmid bands in a 0.7% agarose gel after 5 hr of electrophoresis (40 mA) in Tris-EDTA-boric acid buffer (pH 8). The molecular weight of these plasmid bands were calculated to be: 2.63, 20.95, 5.49, 10.23, 14.12, and 87.10 megadaltons. Plasmids isolated by the sarkosylalkaline denaturation method (Currier and Nester, 1976) show 11 plasmid bands. Our studies indicate that a total number of six plasmid bans are visible in a 014% agarose gel after 20 hr of electrophoresis when plasmid DNA samples are prepared by CsCl-ethidium bromide gradient purification (unpubl.). After several cycles of freezing and thawing of plasmid DNA isolated from CsCl-ethidium bromide gradients, plasmid patterns appear in both CCC and OC forms (Fig. 1). We believe that five bands are OC forms of their counterparts, but we are uncertain of the exact conformation of the bands marked 25 CCC and 10.5 OC. Our values for the molecular weights of CCC forms of B. *thuringiensis* plasmids using plasmid standards with CCC configuration (Macrina et al., 1978) are 3.6, 4.3, 4.8, 10.5, 25.0, and approximately 105 megadaltons. In addition, a substantial number of B. *thuringiensis* var. *israelensis* mutants have been studied in our laboratory which showed a complete loss of parasporal crystals and toxicity against *Aedes aegypti* concomitant with the disappearance of 1 to 3 plasmid bands (Table 2, Fig. 1). We have observed the association of a 4.8 megadalton plasmid with toxic strains and its absence from all non-toxic mutants. It is hoped that further characterization of these mutants will allow us to associate biochemical functions to these plasmids such as antibiotic resistance and bacteriocin production. Once the genetic trait carried in the plasmid is known, it can be confirmed by protoplast transformation of the plasmid DNA into the acrystalliferous mutant strain.

TABLE 2. Bioassay of *Bacillus thuringiensis* var. *israelensis* (BTI) and its mutants using the second instar larvae of *Aedes aegypti*.

BTI Mutants[a]	Spore Formation[b]	Crystal Formation[b]	Toxicity
4Q2-22	+	-	0
4Q2-32	+	-	0
4Q2-40	+	-	0
4Q2-44	+	-	0
4Q2-50	+	-	0
4Q2-52	+	-	0
4Q2-56	+	-	0
4Q2-58	+	-	0
4Q2-60	+	-	0
original *B. thuringiensis* var. *israelensis* (4Q1)	+	+	100%

[a]BTI mutants, except 4Q2-60 which was a spontaneous acrystalliferous dissociate, were isolated by incubating cells of crystalliferous parent on nutrient salt media at 41°C for 48-72 hours.

[b]Cells of BTI or its mutants were inoculated into 100 ml nutrient broth supplemented with $CaCl_2.2H_2O$, $MnCl_2.4H_2O$, and $FeSO_4.7H_2O$ and incubated at 30°C until cell lysis was almost complete. Then the culture was centrifuged and resuspended in 5 ml of sterile distilled water. For identification of spore and crystal formation the cells were withdrawn from the culture after 2 days of growth at 30°C and checked by phase contrast microscopy.

[c]Toxicity was expressed as % mortality. For testing toxicity, 0.5 ml of the spore crystal suspension (20x concentrated) or 10-fold dilutions (10^{-1}, 10^{-2}, 10^{-3}) of the BTI preparation were added to a bioassay cup containing 20 ml of dechlorinated tape water and 20 second-instar larvae of *Aedes aegypti*. Bioassays were incubated at room temperature (23°C). Each dilution was tested in duplicate and mortality was recorded every 24 hr for 72 hr.

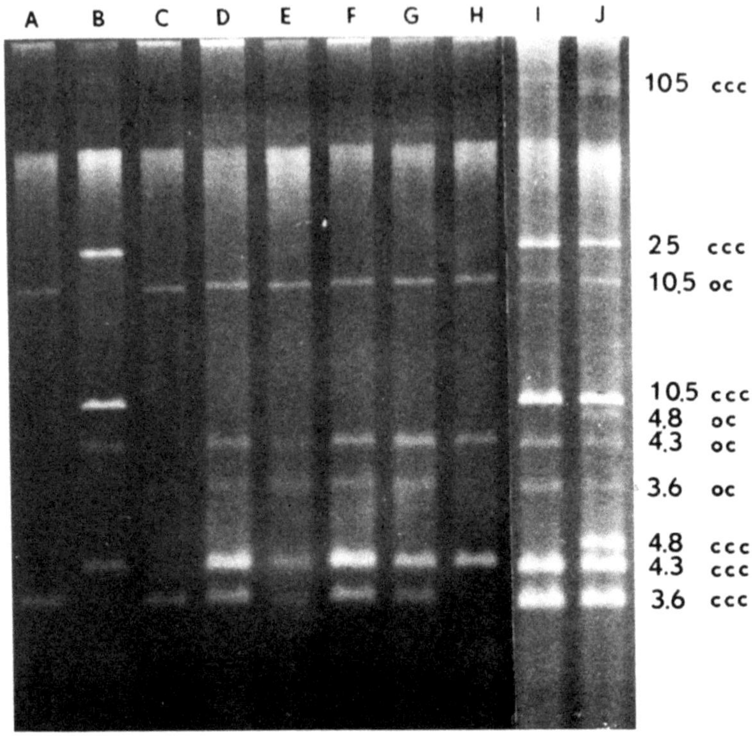

Figure 1. Plasmid profiles of *Bacillus thuringiensis* var. *israelensis* and its mutants.

Plasmid DNA was isolated by Sarkosyl-alkaline denaturation method according to Currier and Nester (1976). Each DNA sample in 40 µl was mixed with 15 µl of tracking dye (0.1M EDTA, 30% sucrose, and 0.0005% bromophenol blue) and applied to a 0.4% agarose gel. Electrophoresis was performed at 20 volts for 20 hr at room temperature in 0.05 M Tris, 0.002 M EDTA, 0.02 M sodium acetate buffer (pH 8.05). The gel was stained with 0.3 µg/ml of ethidium bromide for 30 min. and photographed on a Chromato-Vue transilluminator (Model C-61) using a Polaroid MP-4 land camera with Polaroid type 667 film and a Vivitar No. 25 red filter.

Lane A to I are plasmid profiles of BTI mutants 4Q-22, 4Q2-32, 4Q2-40, 4Q-44, 4Q2-50, 4Q2-52, 4Q2-56, 4W2-58, and 4Q2-60, respectively. Lane J is the plasmid profile of the original BTI (4Q1) which has a total number

V. CONCLUSIONS

The genetics of *B. thuringiensis* is currently at a reasonable working level. Generalized transducing phages CP-51 and CP-54 have proven to be useful in mapping the genome of some varieties, but are hampered by their extremely lytic nature in most varieties. The TP phages hold greater promise for generalized tranduction for *B. thuringiensis* because of their temperate nature.

The ability to transform *B. thuringiensis* is extremely important for a number of genetic studies, including studies on the coding ability of plasmids, genetic mapping, and genetic engineering of insect pathogenesis. Both standard bacterial transformation and protoplast transformation have been reported and will undoubtedly play an important role in future experiments.

The literature on plasmids in *B. thuringiensis* has been extensively reviewed in this paper. Plasmids have been found in almost every variety of *B. thuringiensis* and there have been numerous suggestions that the gene(s) responsible for insect toxicity is borne on a plasmid. There is no direct proof to support this assertion, but there is considerable circumstantial evidence that the crystal genes are plasmid-borne. For example, the crystal is irreversibly lost upon treatment of cells with agents that cure plasmids. On the other hand, curing also results in the high frequency loss of the ability to sporulate. If it is reasonable to conclude that the crystal gene is borne, then it seems reasonable to conclude that some functions associated with sporulation are also plasmid-borne. In fact, Bulla et al. (1977) and Stahly et al. (1978b) have suggested that the parasporal crystal may play a structural role in the spore coat. Selective curing studies reported in this review indicate that one particular

---

Figure 1. (Continued)

of 11 bands, 5 of them are believed to be OC forms of thier counterparts. The OC form of the 25 megadalton plasmid is belived to be immediately beneath the 105 megadalton CCC plasmid. The molecular weights, as expressed in values of megadaltons, were calculated by comparing with CCC DNA molecular weight standards. The smear band seen above the 25-megadalton plasmid band is residual chromosomal DNA.

CCC: closed covalent circle; OC: open circle.

plasmid is present in toxic strains of *B. thuringiensis* var. *israelensis* and absent in non-toxic variants.

Unfortunately the relationship between plasmids and crystal production is not clear for all varieties. The work of Gonzales and Carlton (1980) indicated that no particular plasmid was common to all of the toxic varieties. Furthermore, no clear relation between crystals and plasmids could be seen in the non-toxic strains they studied. The work of Klier et al. (1978) indicates that the crystal gene is expressed as a stable messenger RNA and that this message hybridizes to the chromosomal DNA, not to plasmid DNA. Despite the conflicting reports, the overwhelming body of literature and the general assumption of workers in the field indicates that the toxin is coded for by a plasmid in at least some varieties of *B. thuringiensis*.

VI. REFERENCES

Alikhanian, S. I., Ryabchenko, N. F., Bukanov, N. O., and Sakanyan, V. A. (1981). Transformation of *Bacillus thuringiensis* subsp. *galleria* protoplasts by plasmid pBC 16. *J. Bacteriol.*, 146, 7-9.

Aronson, A. I. (1965). Membrane-bound messenger RNA and polysomes in sporulating bacteria. *J. Mol. Biol.*, 13, 92-104.

deBarjac, H. (1970). Transduction chez *Bacillus thuringiensis*. *C. R. Acad. Sci. Paris*, 270, 2227-2229.

deBarjac, H. (1978a). Une nouvelle variete de *Bacillus thuringiensis* tres toxique pour les Moustiques: *B. thuringiensis* var. *israelensis* serotype 14. *C. R. Acad. Sci. Paris*, 286D, 797-800.

deBarjac, H. (1978b). Toxicite de *Bacillus thuringiensis* var. *israelensis* pour les larves d'*Aedes aegypti* et d'*Anopheles stephensi*. *C. R. Acad. Sci. Paris*, 286D, 1175-1178.

deBarjac, H. and Larget, I. (1979). Proposals for the adoption of a standardized bioassay method for the evaluation of insecticidal formulations derived from serotype H.14 of *Bacillus thuringiensis*. WHO document, WHO/VBC/79.744.

Bibb, M. J., Ward, J. M., and Hopwood, D. A. (1978). Transformation of plasmid DNA into *Streptomyces* at high frequency. *Nature*, 274, 388-400.

Bulla, L. A., Jr., Kramer, K. J., and Davidson, L. I. (1977). A sporulation specific protein of *B. thuringiensis*. *Abstr. Annu. Meet. Am. Soc. Microbiol.*, I 62, p. 165.

Chang, S. and Cohen, S. N. (1979). High frequency transformation of *Bacillus subtilis* protoplasts by plasmid DNA. *Mol. Gen. Genet.*, 168, 111.115.

Currier, T. C. and Nester, E. W. (1976). Isolation of covalently closed circular DNA of high molecular weight from bacteria. *Anal. Biochem.*, 76, 431-441.

Debabov, V. G., Azizbekyan, R. R., Khlebalina, O. I., D'Yachenko, V. V., Galushra, F. P., and Belykh, R. A. (1977). Isolation and preliminary characterization of extrachromosomal elements of *Bacillus thuringiensis* DNA. *Genetica*, 13, 496-501.

Delafield, F., Somerville, H. J., and Rittenberg, S. C. (1968). Immunological homology between crystal and spore protein of *Bacillus thuringiensis*. *J. Bacteriol.*, 96, 713-720.

Ermakova, L. M., Galushka, F. P., Strongin, A. Y., Sladkova, I. A., Rebentish, B. A., Andreeva, M. V., and Stepanov, V. M. (1978). Plasmids of crystal forming *Bacillus* and the influence of growth medium composition on their appearance. *J. Gen. Microbiol.*, 107, 169-171.

Faust, R. M., Spizizen, J., Gage, V. and Travers, R. S. (1979). Extrachromosomal DNA in *Bacillus thuringiensis* var. *kurstaki*, var. *finitimus*, var. *sotto*, and in *Bacillus popilliae*. *J. Invertebr. Pathol.*, 33, 233-238.

Fitz-James, P. C. and Young, I. E. (1959). Comparison of species and varieties of the genus *Bacillus*. *J. Bacteriol.*, 78, 743-754.

Glatron, M. R. and Rapoport, G. (1972). Biosynthesis of the parasporal inclusion of *Bacillus thuringiensis:* Half-life of its corresponding messenger RNA. *Biochimie*, 54, 1291-1301.

Glatron, M. R. and Rapoport, G. (1975). In vivo and in vitro evidence for existence of stable messenger ribonucleic acids in sporulating cells of *Bacillus thuringiensis*. *In* "Spores IV" (P. Gerhardt, R. N. Costilo, and H. L. Sadoff, ed.). American Society for Microbiology, Washington, D.C. pp. 255-264.

Goldberg, L. J. and Margalit, J. (1977). A bacterial spore demonstrating rapid larvicidal activity against *Anopheles seogento, Uranotaenia unguiculata, Culex univitattus, Aedes aegypti,* and *Culex pipiens*. *Mosquito News*, 37, 355-358.

Gonzalez, J. M. and Carlton, B. C. (1980). Patterns of plasmid DNA in crystalliferous and acrystalliferous strains of *Bacillus thuringiensis*. *Plasmid*, 3, 92-98.

Iizuka, T., Faust, R. M., and Travers, R. S. (1981a). Isolation and partial characterization of extrachromosomal DNA from serotypes of *Bacillus thuringiensis* pathogenic to lepidopteran and dipteran larval by agarose gel electrophoresis. *J. Sericult. Sci. Japan*, 50, 144.

Iizuka, T., Faust, R. M., and Travers, R. S. (1981b). Comparative profiles of extrachromosomal DNA in single and multiple crystalliferous strains of *Bacillus thuringiensis* var. *kurataki*. *J. Fac. Agr. Hikkaido Univ. Japan*, 60, 143-151.

Klier, A. and Lecadet, M. M. (1976). Arguments based on hybridization-competition experiments in favor of the *in vitro* synthesis of sporulation-specific mRNAs by the RNA polymerase of *B. thuringiensis*. *Biochem. Biophys. Res. Comm.*, 73, 263-270.

Klier, A., Lecadet, M. M., and Rapoport, G. (1978). Transcription in vitro of sporulation specific RNAs by RNA polymerase from *Bacillus thuringiensis*. *In* "Spores VII" (G. Chambliss and J. C. Vary, ed.). The American Society of Microbiology, Washington, D.C. pp. 205-212.

Krieg, A., deBarjac, H., and Bonnefoi, A. (1968). A new serotype of *Bacillus thuringiensis* isolated in Germany: *Bacillus thuringiensis* var. *darmstadiensis*. *J. Invertebr. Pathol.*, 10, 428-430.

Lacey, R. W. (1975). Antibiotic resistance plasmids of *Staphylococcus aureus* and their clinical importance. *Bact. Rev.*, 39, 1-32.

Lecadet, M. M., Blondel, M. O., and Ribier, J. (1980). Generalized transduction in *Bacillus thuringiensis* var. Berliner-1715 using bacteriophage CP-54Ber. *J. Gen. Microbiol.*, 121, 203-212.

Lovett, P. S. and Bramucci, M. G. (1975). Plasmid deoxyribonucleic acid in *Bacillus subtilis* and *Bacillus pumilus*. *J. Bacteriol.*, 124, 484-490.

Lüthy, P. (1980). Insecticidal toxins of *Bacillus thuringiensis*. *FEMS Microbiol. Let.*, 8, 1-7.

Macrina, F. L., Kopecko, D. J., Jones, K. R., Ayers, D. J., and McCowen, S. M. (1978). A multiple plasmid-containing *Escherichia coli* strain: convenient source of size reference plasmid molecules. *Plasmid*, 1, 407-410

Martin, P. A. W. (1979). A genetic system for *Bacillus thuringiensis*. *Doctoral Dissertation*. The Ohio State University. Columbus, Ohio. 116 pp.

Martin, P. A. W. and Dean, D. H. (1980). *In* "Plasmids and Transposons" (C. Stuttard and K. Rozee, ed.). Academic Press, New York. pp. 155-161.

Martin, P. A. W., Lohr, J. R., and Dean, D. H. (1981). Transformation of *Bacillus thuringiensis* protoplasts by plasmid deoxyribonucleic acid. *J. Bacteriol.*, 145, 980-983.

Miteva, V. I. (1978). Isolation of plasmid DNA from various strains of *Bacillus thuringiensis* and *Bacillus cereus*. *Doklady Bolgarskoi Akademii Nauk.*, 31, 913-916.

Nickerson, K. W. and Bulla, L. A. (1974). Physiology of spore-forming bacteria associated with insects: minimal nutritional requirements for growth, sporulation, and parasporal crystal formation of *Bacillus thuringiensis*. *App. Microbiol.*, 28, 124-132.

Norris, J. R. (1970). Sporeformers as insecticides. *J. Appl. Bacteriol.*, 33, 192-206.

Perlak, F. J., Mendelsohn, C. L., and Thorne, C. B. (1979). Coverting bacteriophage for sporulation and crystal formation in *Bacillus thuringiensis*. *J. Bacteriol.*, 140, 699-706.

Rain-Guion, M. C., Glatron, M. F., Klier, A., Lecadet, M. M., and Rapoport, G. (1976). Coding capacity of the transcription products synthesized *in vitro* by the RNA polymerases from *Bacillus thuringiensis*. *Biochem. Biphys. Res. Comm.*, 70, 709-716.

Reeves, E. L. (1966). Transformation of the entomophilic *Bacillus finitimus* Heimpel and Angus by the DNA from the entomophilic *Bacillus thuringiensis* Berliner. *Doctoral Dissertation*. The Ohio State University. Columbus, Ohio. pp. 52.

Reeves, E. L. and Garcia, C. (1971). Pathogenicity of bicrystalliferous *Bacillus* isolated from *Aedes aegypti* and other aedine mosquito larvae. *In* "International Colloquium on Insect Pathology", College Park, MD. August 1970, pp. 219-228.

Ryabchenko, N. F., Bukaov, N. O., Sakanian, V., and Alikhanian, S. I. (1980). Transformation of protoplasts of *Bacillus thuringiensis* var. *galleria* 69-6 by plasmid PBC16. *Dokl. Akad. Nauk. SSSR*, 253, 729-732.

Ribier, J. and Lecadet, M. M. (1973). Etude ultrastructurale et cinetique de la sporulation of *Bacillus thuringiensis* var. *Berliner* 1715. Remarques sur la formation de l'inclusion parasporale. *Ann. Inst. Pasteur (Paris)*, 124A, 311-344.

Somerville, H. J., Delafield, F. P., and Rittenberg, S. C. (1968). Biochemical homology between crystal and spore protein of *Bacillus thuringiensis*. *J. Bacteriol.*, 96, 721-726.

Spizizen, J. (1958). Transformation of biochemically deficient strains of *Bacillus subtilis* by deoxyribonucleate. *Proc. Nat. Acad. Sci.*, 44, 1072-1078.

Spizizen, J. (1959). Genetic activity of deoxyribonucleic acid in the reconstitution of biosynthetic pathways. *Fed. Proc.*, 18, 957-965.

Spizizen, J. (1961). Studies on transformation of sporulating characters. *In* "Spores II" (H. O. Halvorson, ed.). The American Microbiology, Washington, D.C. pp. 142-148.

Stahly, D. P., Dingman, D. W., Field, C. C., Feiss, M. G., and Smith, G. L. (1978a). Multiple extrachromosomal deoxyribonucleic acid molecules in *Bacillus thuringiensis*. *FEMS Microbiol. Let.*, 3, 139-141.

Stahly, D. P., Dingman, D. W., Bulla, L. A., and Aronson, A. I. (1978b). Possible origin and function of the parasporal crystals in *Bacillus thuringiensis*. *Biochem. Biophys. Res. Comm.*, 84, 581-588.

Steinhaus, E. A. (1975). *In* "Disease in a Minor Chord". Ohio State Univ. Press, Columbus, Ohio. 488 pp.

Tanaka, T., Kuroda, M., and Sakaguchi, K. (1977) Isolation and characterization of four plasmids from *B. subtilis*. *J. Bacteriol.*, 129, 1487-1494.

Thorne, C. B. (1968a). Transducing bacteriophage for *Bacillus cereus*. *J. Virol.*, 2, 657-662.

Thorne, C. B. (1968b). Transduction in *Bacillus cereus* and *Bacillus anthracis*. *Bacteriol. Rev.*, 32, 358-361.

Thorne, C. B. (1978). Transduction in *Bacillus thruingiensis*. *Appl. Envir. Microbiol.*, 35, 1109-1115.

Toumanoff, C. (1955). Au sujet de souches cristallophores entomophytes de *Bacillus cereus*, observations sur leurs inclusions cristallines. *Annal L'Institut Pasteur*, 89, 644-653.

Undeen, A. H. and Colbo, M. H. 1980. The efficacy of *Bacillus thuringiensis* var. *israelensis* against blackfly larvae (Diptera: Simuliidae) in their natural habitat. *Mosquito News*, 40, 181-184.

Van Essen, F. W. and Hembree, S. C. (1980). Laboratory bioassay of *Bacillus thuringiensis israelensis* against all instars of *Aedes aegypti* and *Aedes taeniorhynchus* larvae. *Mosquito News*, 40, 424-431.

Vankova, J. (1957). Study of the effect of *Bacillus thuringiensis* on insects. *Folia Biologica*, 3, 175-183.

Yelton, D. B. and Thorne, C. B. (1970). Transduction in *Bacillus cereus* by each of two bacteriophages. *J. Bacteriol.*, 102, 573-579.

Yousten, A. A., Hanson, R. S., Bulla, L. A., and St. Julian, G. (1978). Physiology of sporeforming bacteria associated with insects. V. Tricarboxylic acid cycle activity and adenosine triphosphate levels in *Bacillus popillae* and *Bacillus thuringiensis*. *Canad. J. Microbiol.*, 20, 1729-1724.

Zakharyan, R. A., Agabalyan, A. S., Chil-Akopya, L. A., Gasparyan, N. C., Bakunts, K. A., Tatevosyan, P. E., Afrikyan, E. K. (1976). Possible role of extrachromosomal DNA in the formation of the entomocidal endotoxin of *Bacillus thuringinensis*. *Dokl. Akad. Nauk*, ArmSSR., 63, 42-47.

Zakharyan, R. A., Israelyan, Y. A., Agabalyan, A. S., Tatevosyan, P. E., Akopyan, S. M., Afrikyan, E. K. (1979). Plasmid DNA from *Bacillus thuringiensis*. *Mikrobiol.*, 48, 226-229.

STATUS OF RESEARCH ON BIOLOGICAL AGENTS FOR THE CONTROL OF MOSQUITOES

D.W. Anthony

Insects Affecting Man and Animals Research Laboratory
Agricultural Research
Science and Education Administration
U.S. Department of Agriculture
Gainesville, Florida

I. INTRODUCTION

Mosquitoes are one of the most important groups of bloodsucking arthropods that annoy and transmit diseases to man and livestock. They are vectors of viral, protozoan, and filarial diseases, and they are the sole vectors of the Group A arboviruses, which include Venezuelan equine encephalitis, Western equine encephalitis, and Eastern equine encephalitis. Human and avian malarias (*Plasmodium* spp.) are transmitted exclusively by mosquitoes, and they are the principal transmitters of filariasis to man and other animals. Also, from the standpoint of veterinary medicine, the annoyance factor brought about by high density mosquito populations cannot be overstated. Large areas of irrigated farmland and vast marshes along sea coasts frequently produce very high mosquito populations thus retarding livestock production and further agricultural development. Steelman et al. (1972, 1973) have reported that high density populations of *Anopheles quadrimaculatus*, *A. crucians*, and *Psorophora columbiae (=P. confinnis)* can cause significant reductions in the average daily weight gain of feedlot steers.

During the last 20 years a great deal of emphasis has been placed on the use of parasites, predators, and pathogens as biological agents for the control of insect pests of crops and forests. However, only recently has there been any appreciable effort to develop biocontrol technologies for use against mosquitoes and other arthropods of veterinary and medical importance. Although chemicals are still (and probably will remain for the foreseeable future) the principal agents for mosquito control, noninsecticidal methods of control are badly needed as adjuncts to or replacements for certain insecticides. The number of available pesticides is being reduced each year by either the occurrence of resistance in target species or by restrictions on their use in the environment. Researchers and control specialists agree that sound control programs cannot depend upon a single method of control, but must incorporate all appropriate methods. Biological control agents, such as viral or microbial pathogens or parasites, will be essential and integral parts of integrated control schemes.

Although biological control of mosquitoes may include autocidal (sterile male or genetic manipulation) techniques, pheromones, growth inhibitors, source reduction methods such as water management, as well as predators (vertebrates and invertebrates), this discussion will be limited to those agents generally regarded as pathogens, i.e., the viruses, bacteria, protozoa, fungi, and nematodes. Chapman (1974) summarized in detail most of the early and present day knowledge concerning the biological control of mosquito larvae, and this report will make no attempt to reiterate the bulk of this information. Instead, an effort will be made to update the excellent report by Chapman in light of more recent studies and to emphasize research with some of the more promising biological agents.

II. VIRUS DISEASES

Federici (1977) listed 130 known virus-mosquito host records. These records included the nuclear polyhedrosis viruses (NPVs *Baculoviruses*), the cytoplasmic polyhedrosis viruses (CPVs), the iridescent viruses (MIVs, *Iridoviruses*), a densonucleosis virus (*Parvovirus*) an *Entomopoxvirus* and several unclassified non-occluded viruses. Also, in laboratory tests, at least two iridescent viruses from lepidopteran hosts have been transmitted to mosquito larvae (Fukuda, 1971; Anthony, unpubl.). Table 1 lists some of the genera of mosquitoes that are of veterinary importance and the virus diseases recorded from at least one species of that genus.

Nuclear Polyhedrosis Viruses. These viruses infect the nuclei of the epithelial cells of the midgut and gastric caeca, and the infections are nearly always fatal to larvae. Although the nuclear polyhedrosis viruses (NPVs) have been found in at least 11 species

TABLE 1. Some genera of mosquitoes of veterinary Importance and their viruses.

Mosquito Genera	NPVs[a]	CPVs[b]	MIVs[c]	Densonucleosis	Entomopox	Other
Aedes	+	+	+	+	+	+[d,f]
Anopheles	+	+				+[e]
Culex	+	+				+[e,f]
Culiseta		+				+[d]
Psorophora	+	+	+			

[a]Nuclear polyhedrosis viruses.
[b]Cytoplasmic polyhedrosis viruses.
[c]Mosquito iridescent viruses.
[d]Non-occluded virus (Federici, 1973).
[e]Includes a tetragonal virus (Kellen et al., 1963) and/or a non-occluded virus in adults (Davies et al., 1971).
[f]Tetragonal virus (Kellen et al., 1963; Clark and Chapman, 1969).

of mosquito larvae, field epizootics have been seen only in *Aedes sollicitans* (see Clark and Fukuda, 1971). The inability to achieve high infection rates in test larvae and thereby produce large quantities of virus for experimental use has hampered research efforts.

Cytoplasmic Polyhedrosis Viruses. The cytoplasmic polyhedrosis viruses (CPVs) are known to infect many species of mosquitoes. Although the cytoplasm of the midgut, gastric caeca, and occasionally the hindgut epithelial cells may be heavily infected, these viruses usually cause mortality in the host larvae or reductions in adult emergence. However, the effect of these infections on adult longevity or fecundity has not been studied.

Mosquito Iridescent Viruses. The mosquito iridescent viruses (MIVs) have received more attention than any of the other viruses infecting mosquitoes. Among these were some of the first viruses reported infecting mosquitoes and they can be maintained in the laboratory, especially in *Aedes taeniorhynchus* which is easily colonized. MIVs infect the cytoplasm of fat body cells, tracheal epithelium, imaginal bud, nerve, and gonadal tissue. The virus forms paracrystalline arrays in the fat body, thus imparting a characteristic iridescence when a patently infected specimen is viewed against a dark background with reflected light. All patently infected larvae die, usually just before pupation; however, some larvae apparently do not become visibly infected, and these emerge as adults and transmit the infection to their progeny (Linley and Nielson, 1968; Hall and Anthony, 1971). There are 13 world-wide host records of MIVs, and all hosts belong in the genera *Aedes* and *Psorophora*, which possess diapausing eggs. The type of egg may be a very important factor since transovarial transmission of MIVs is common, and diapausing infected eggs are necessary for the survival of the virus during periods of drought and cold. Research efforts with MIVs are now minimal since it has not been possible to obtain high levels of infection in laboratory tests regardless of the dosage administered to test larvae.

"Tetragonal" Viruses. One of the first virus diseases recorded from mosquitoes was thought to be a "possible polyhedrosis virus" (Kellen et al., 1963, 1966) because of the tetragonal shaped crystals observed in infected cells. However, ultrastructure studies by Stoltz et al. (1974) indicate that the crystals observed by light microscopy are not proteinaceous bodies containing virions as is the case with NPVs or CPVs. Instead the electron micrographs show crystals composed of very small particles in hexagonal and recti- linear arrays, which appear to develop in the nucleus of imaginal disc cells and epidermal cells and then invade the cytoplasm. Unequivocal proof of the viral nature of the particle is lacking. However, the disease is transmissible, and infected larvae usually

die. The disease was first observed in larvae of *Culex tarsalis*, *Aedes sierrensis*, and *Anopheles freeborni* (Kellen et al., 1963) and then in other species of *Culex* and *Anopheles* (Clark and Chapman, 1969; Chapman et al., 1970).

Other Viruses. Little is known regarding the densonucleosis and the entomopoxvirus reported from mosquito larvae (Lebedeva and Zelenko, 1972; Lebedeva et al., 1973). Light and electron microscope studies indicate that the viruses observed were similar to densonucleosis and entomopoxviruses in other insects. Similarly, non-occluded virus particles have been reported from larval and adult mosquitoes (Federici, 1977), but apparently little is known of their pathogenicity.

III. BACTERIAL DISEASES

The vast reservoir knowledge that exists from research in other fields of bacteriology provides a platform, perhaps unequaled for any other group of microbial agents, from which to launch into isolating, testing, and the eventual commerical production of entomogenous bacteria pathogenic to mosquitoes. Although many species of bacteria have been isolated from mosquitoes (Singer, 1977), in most cases information regarding their effect on mosquito larvae or adults is lacking. At this time, however, there is a great deal of interest in the spore-forming bacteria as mosquito pathogens, particularly new isolates of *Bacillus thuringiensis* and *B. sphaericus*.

Early efforts to utilize *B. thuringiensis* as a mosquito larvicide were disappointing (Liles and Dunn, 1959; Kellen and Lewallen, 1960). Hall et al. (1977) evaluated formulations from 127 different strains of *B. thuringiensis* against *Aedes* spp. and *Culex* spp. and found that a number of the strains showed significant activity. Goldberg and Margalit (1977) reported a newly isolated *B. thuringiensis* strain that demonstrated rapid larvicidal activity against selected species of *Aedes*, *Anopheles*, *Culex*, and *Uranotaenia*. This newly isolated bacterium was later described as *Bacillus thuringiensis* var. *israeliensis* by de Barjac (1978a). In subsequent studies, de Barjac (1978b) confirmed the high larvicidal activity of this isolate against *Aedes aegypti* and *Anopheles stephensi*, and also pointed out that unlike certain other serotypes of *B. thuringiensis*, this strain was not pathogenic to tested lepidopteran larvae. This new isolate of *B. thuringiensis* is now being studied intensively by mosquito control specialists throughout the world.

Kellen et al. (1965) first showed *B. sphaericus* to be a promising pathogen of mosquitoes. In laboratory tests, ten species of mosquitoes were found to be susceptible to the bacterium, but field tests failed to provide significant reductions in *Aedes* and *Culiseta* larval populations. Recent isolates (Singer, 1973) have shown as much as 10,000 times more insecticidal activity than the strain isolated by Kellen and co-workers. Singer (1975) reported that the SSII-1 strain of *B. sphaericus* isolated from India was more active against *Culex* spp. than against *Anopheles* spp. or *Aedes aegypti*; however, other strains isolated from Indonesia and the Philippines appeared to be more active against anophelines than was the SSII-1 strain, and these strains were also highly active against *Culex* spp. (Singer and Murphy, 1976). In field applications of three strains of *B. sphaericus* reported by Ramoska et al. (1978), natural larval populations of *C. nigripalpus* and *Psorophora columbiae* were reduced by nearly 90% in three experiments. At the Gainesville, Florida, laboratory, we have tested a commercially produced wettable powder formulation of *B. sphaericus* and found that it was highly effective against early instar larvae of *Culex pipiens quinquefasciatus*, *C. salinarius*, and *C. tarsalis*. *Anopheles albimanus* and *A. quadrimaculatus* were intermediate in susceptibility while *Aedes aegypti* and *A. taeniorhynchus* showed low mortalities even at very high dosages.

Manufacturers of *B. thuringiensis* preparations for the control of crop and forest pests have shown great interest in *B. thuringiensis* var. *israeliensis* and *B. sphaericus* since both appear to be promising as mosquito larvicides. Although there is still much developmental work to be done, it seems probable that one or both of these new "biologicals" may be commercially available in the near future.

IV. PROTOZOAN DISEASES

Numerous protozoan species have been identified from mosquitoes. Many of these, especially the external ciliates (*Vorticella* and *Epistylis*), internal ciliates (*Tetrahymena* and *Lambornella*), flagellates (*Crithidia*), eugregarines (*Lankesteria*), and schizogregarines (*Caulleryella*) have been insufficiently studied because of their low pathogenicity.

Microsporidia are common parasites of nearly all arthropods, and mosquitoes are not exception. Almost all mosquitoes have diseases caused by the microsporidian genera *Amblyospora*, *Hyalinocysta*, *Parathelohania*, and *Pilosporella* (these genera all belong to the family Thelohaniidae) (Hazard and Oldacre, 1975). *Amblyospora* and *Hyalinocysta* infect species of *Aedes*, *Coquilletidia*, *Culex*, *Culiseta*, *Mansonia*, and *Psorophora*; *Parathelohania* are primarily parasites of *Anopheles* spp. *Pilosporella* spp. are restricted to mosquitoes breeding in bromeliads and artificial containers (*Wyeomyia* and a few species of *Aedes*). At the present time, 84 species of mosquitoes have been reported as hosts of these microsporidians (Hazard, 1977; Hazard and Chapman, 1977).

All species of the family Thelohaniidae have a highly specialized sequence of development in the adult female mosquito that invariably results in transovarial (vertical) transmission to most of the progeny and to succeeding generations. The infection is benign in females, but in most cases it kills all male larvae. Although millions of spores are formed in these male larvae, they are not infectious per os to their hosts, a situation that has puzzled many investigators for years. Recent studies at the U.S. Department of Agriculture's Gainesville, Florida, laboratory have shown that these infections cannot persist in natural mosquito populations through transovarial transmission alone. Cytogenetic studies have demonstrated that sporonts of *Amblyospora*, *Parathelohania*, and possibly other genera undergo meiotic divisions that give rise to haploid spores (Hazard et al., 1979). This phenomenon highlights the difference between these vertically transmitted microspordia and those microsporidia that are readily transmitted per os (horizontal transmission) such as *Nosema algerae* and *Vavraia culicis*. The information obtained from the cytogenetic studies suggests a sexual cycle (gametogamy), perhaps in an alternate host, since these haploid spores are not infectious to their mosquito hosts. Computer models, assuming that we can learn how to utilize per os transmission of these microsporidia, indicate great potential for these diseases as biological control agents. Some of these agents that have received the most attention are *Nosema algerae*, *Vavraia culicus*, *Hazardia milleri*, and *Helicosporida* spp.

Personnel at the Gainesville laboratory have spent considerable time and effort studying the effects of *N. algerae* on anopheline mosquitoes (particularly *A. albimanus*), and efforts have been made to develop methodology for its practical use in anopheline breeding areas. Although this microsporidian has been reported to be highly pathogenic to larvae of some species of *Anopheles*, its principal effect on *A. albimanus* results in a reduction of longevity and fecundity in adult females (Anthony et al., 1972). Tests have shown that infected females live only about one-half as long as uninfected females and that the infected produce an average of 54% fewer viable eggs (Anthony et al., 1978). These reductions are highly significant in regard to disease transmission and population reduction aspects.

Nosema algerae also produces pathogenic effects in *A. quadrimaculatus*, *C. nigripalpus*, and *tarsalis*. I_{100} dosages (number of spores required to produce 100% infection) are slightly higher for these species than required for *A. albimanus*. The infection reduces the longevity of *A. quadrimaculatus* and *C.*

tarsalis; however, its relationship to fecundity has not been thoroughly studied (Anthony, unpubl.).

Spores of *N. algerae* can be mass produced in *Heliothis zea* with an average yield of about one billion spores per infected specimen. The cost of producing sufficient spores to treat 1 hectare (2.5 acres) of anopheline breeding area at a rate of 1 X 10^7 spores/m^2 is about $9.00. This estimate is based on the cost of time and materials only and does not include capital investment such as buildings, remodeling, or necessary equipment. Admittedly, the cost is not in the range of economic feasibility at this time; however, increased yields in production or improved formulation techniques that effectively lower required dosages could reduce this figure significantly. Improved methods of harvesting, clearing, and storing spores also have been developed. Aqueous spore suspensions that are relatively free of host tissue and debris have been stored for 6 months at 4-6°C without a measurable loss in infectivity.

Field tests in which *N. algerae* was used against *A. albimanus* in the Panama Canal Zone resulted in infections of *A. albimanus* in all test plots (Anthony et al., 1978). Infection rates were dose-dependent and ranged from 16% in a test area of 1,100 m^2 treated with a single application of 2.15 X 10^7 spores/m^2 to 86% in a test plot of 37 m^2 that were treated four times with 2.15 X 10^9 spores/m^2. Also, many infected larvae brought to the laboratory died before pupation, and microscopical examinations showed that these larvae were heavily infected with *N. algerae*. These tests indicate that applications of spores of *N. algarae* to natural breeding areas can produce infections in native populations of *A. albimanus*.

Vavraia culicis, previously known as *Pleistophora culicis,* is readily transmitted per os to a wide range of mosquito species (Weiser and Coluzzi, 1972). Recent studies in our laboratory have shown high infection rates in *A. albimanus, A. taeniorhynchus, C. salinarius,* and *C. tarsalis.* Histological studies of infected adults of *A. albimanus* show that the Malpighian tubules are the principal sites of infection but there is some involvement of fat body,gut, and muscle tissues. Infected females die sooner than those mosquitoes that are not infected, especially if they are given blood meals. We have also found that *V. culicis* can be mass produced in *H. zea* at a rate of about 300 million to 500 million spores per infected specimen. The technique are identifical to those used for the production of *N. algerae* (Kelly and Anthony, unpubl.). This improvement in spore production will for the first time allow the full evaluation of this microsporidian as a potential biocontrol agent for several species of mosquitoes.

Hazardia milleri, originally described as *Stempellia milleri* by Hazard and Fukuda (1974), is primarily infectious to larvae of *C. pipiens quinquefasciatus*. However, per os infections have also been obtained in *C. resturans* and in *C. salinarius*. Infected larvae usually die before pupation; however, studies have been hampered by difficulties in obtaining high infection rates in laboratory tests.

Recently, *Helicosporidum parasiticum* was isolated from field collected larvae of *C. nigripalpus* and was shown to be infective to 14 mosquito species in six genera. However, high dosages were required to produce substantial infections (Fukuda et al., 1976).

V. FUNGI

Many species of fungi have been recorded from mosquitoes (Roberts, 1977), but most of these have not received sufficient study to be evaluated as mosquito pathogens. The genera containing the most important pathogens of mosquites are *Coelomomyces*, *Lagenidium*, *Culicinomyces*, *Entomophthora*, *Beauveria*, and *Metarrhizium*.

Coelomomyces. Fourteen species of this fungal genus, all apparently specific to mosquitoes, are known in the United States. Twenty-seven mosquito in 8 genera have been found infected with one or more of these fungal species. McNitt and Couch (1977) have listed the world-wide distribution of species and host records.

Some species of *Coelomomyces* are very persistent in field situations (Umphlett, 1970; Chapman and Glenn, 1972), but natural infection levels may be extremely variable. Field infections have been induced by several investigators (Laird, 1967; Couch, 1972; Chapman, 1974) by the dissemination of infected larval cadavers. These positive results were indeed fortunate since the complete life cycle of *Coelomomyces* was not known at that time. Whisler et al. (1974, 1975) found that the copepod (*Cyclops vernalis*) was an intermediate hosts of a species of *Coelomomyces*, and other workers have since corroborated this highly significant achievement (Federici, 1975; Federici and Chapman, 1977). We now know that a microcrustacean is essential for disease transmission to mosquitoes and at least four species of *Coelomomyces* have been maintained in various laboratories. Field releases await safety tests and decisions on how the fungus can best be produced and disseminated.

Lagenidium. *Lagenidium giganteum* infects the larvae of many mosquito species. The fungus can be cultured by using a variety of artificial media, or it can be maintained in mosquito larvae. The staff at the Gulf Coast Mosquito Research Laboratory, Lake Charles, Louisiana, has followed natural epizootics of this fungus in *Culex territans* in a swamp for 3 years. These studies showed that the fungus produced extremely high levels of infection (H. C. Chapman, pers. commun.). McCray et al. (1973) conducted limited small plot field tests with variable results; however, in some tests high infection levels were obtained. Improved mass production and formulation technology as well as safety tests are now necessary.

Culicinomyces. Couch et al. (1974) described *Culicinomyces clavosporus* from larvae of *A. quadrimaculatus.* The fungus can be cultured on artificial media. Infections are initiated by the ingestion of conidia, and death occurs within 60 hr after infection. Larvae of ten other species of mosquitoes were reported to be susceptible to the fungus.

Entomophthora. Species of *Entomophthora* are known to produce fatal infections in several species of adult mosquitoes. A single species, *E. aquatica,* was found infecting larvae and pupae of *Aedes canadensis* and *Culiseta morsitans* (see Anderson and Ringo, 1969). Roberts (1974) indicates that although high infection levels of adults have been noted in nature, artificial introductions of these fungi into natural breeding or overwintering sites have not been attempted.

Beauveria. Beauveria bassiana is one of the most frequently isolated entomogenous fungi, although it is rarely found in natural populations of mosquitoes. Laboratory and field tests by Clark et al. (1968) indicated that *A. albimanus, C. pipiens,* and *C. tarsalis* were susceptible to *B. bassiana.* However, two species of *Aedes* were not susceptible. In small scale outdoor tests with conidia of this species, Clark and co-workers recorded significant reductions in larvae and pupae of *C. pipiens;* but the dosage used was equivalent to 3 lb of conidia per acre. The extremely high required dosage and the apparent nonsusceptability of *Aedes* larvae led these investigators to conclude that the use of *B. bassiana* conidia for control of mosquito larvae was not practical.

Metarrhizium. Roberts (1974) has reviewed in detail the potential use of *Metarrhizium anisopliae* as a microbial insecticide for mosquito control. This fungus has a very wide host range, but it has not been isolated in nature from aquatic insects. Tests have shown that larvae of at least 11 species of mosquitoes, all members of the major genera of veterinary and medical importance, are susceptible to the conidia. The fungus can be mass produced on artificial media, but there is evidence that the medium in which the conidia are produced may affect the virulence. Small plot outdoor tests have resulted in high reductions of *C. pipiens pipiens* and *A. sollicitans.* Roberts (1974) suggests that *M. anisopliae* may be useful in brackish water and in water that is organically polluted, i.e., breeding areas that may be unsuitable for certain other mosquito pathogens. In laboratory tests and in field tests, Roberts (1975) found that *Anopheles gambiae* larvae were susceptible to *M. anisopliae* conidia. Small plot tests in muddy water and clear water situations showed that 100% mortality of larger larvae and pupae was obtained when dry aliquots of conidia were dusted on the water surface at the rate of 600 mg of conidia per m^2. Population reduction was maintained longer in clear water than in muddy water. *Anopheles gambiae* was the principal test species, but some *Anopheles rufipes* and *Anopheles funestus* occurred at the clear water test sites.

VI. MERMITHID NEMATODES

Mermithid nematodes appear to meet virtually all the criteria for the ideal biological control agent: They have adapted to their host's life cycle, they kill their hosts, they produce high incidences of infection, and they have a high reproductive potential. Some mermithid nematodes can be handled and mass produced in the laboratory. They are easily disseminated, they have the potential for establishment and recycling in the breeding area, and they offer no threat to nontarget organisms or to the environment.

Although a number of genera and species of mermithids have been studed to varying degrees, this report will discuss only the two that have received the most attention, i.e., *Diximermis peterseni* and *Romanomermis culicivorax*.

Diximermis peterseni. This species is host specific for *Anopheles* spp. and it is easily maintained in the laboratory (Peterson and Chapman, 1970). Peterson and Willis (1974b) and Woodard (1978) have reported small-scale field releases that showed *D. peterseni* has a strong potential for establishment and recycling in anopheline larval habitats. Very high levels of infection were obtained in some tests. Because of the difficulty of mass rearing *A. quadrimaculatus* (the laboratory host), production of sufficient quantities of this nematode for extensive field testing has not been possible.

Romanomermis culicivorax. This species (formerly referred in some reports as *Romanomermis (Reesimermis nielseni)* has been found to parasitize 17 species of mosquitoes in six genera in nature, and it is known to develop in about 85 species in the laboratory. The biology of *R. culicivorax* has been studied extensively in the laboratory and in the field (Petersen, 1972; 1973a,b), and much of the knowledge concerning mermithids has been obtained directly or indirectly from studies of this species.

Petersen and Willis (1972) have successfully released *R. culicivorax* by introducing cultures of the preparasitic stages directly into mosquito breeding areas with standard compressed-air hand sprayers or from a helicopter with standard spray equipment. Numerous small scale releases (Petersen and Willis, 1974a) produced levels of infection that averaged 85% in *Anopheles* mosquitoes at dosage rates of 2,000 preparasites/yd^2 (0.84 m^2) of surface area. Recycling has been observed in a number of the release sites with infection levels reaching 90-100% 4 years after introduction of the nematode. Also, significant levels of parasitism of *Aedes* spp. and *Psorophora* spp. were obtained over a period of 18 weeks and six floodings following introductions of the nematode as a prehatch treatment.

Mammalian safety tests and tests against nontarget organisms have shown that *R. culicivorax* offers no threat to warm-blooded animals or to the environment (Ignoffo et al., 1973, 1974). Because of its highly complex and specialized nature, it offers not potential hazard of undergoing undesirable changes or modification after release. Also, it does not threaten competitive displacement of other desirable organisms.

R. culicivorax is the only mermithid that has been mass produced in sufficient quantities to permit large scale field release studies (Petersen et al., 1978a). This asset has led to the most extensive study to date for the control of mosquitoes with pathogens or parasites. Petersen et al. (1978b) released *R. culicivorax* 11 times over a 1.2 hectaries of anopheline breeding areas in Lake Apastepeque, El Salvador. Although parasitism averaged only 58%, *Anopheles* populations were reduced by 94%. These highly encouraging results suggest that *R. culicivorax* is indeed a highly effective agent and can be incorporated into management schemes for the control of mosquitoes.

VII. CONCLUSIONS

During the last ten years there has been expanded research on biological agents for the control of mosquitoes. Encouraging results have been obtained with mermithid nematodes, bacterial pathogens, and certain protozoans (Microsporidia). The nematode, *Romanomermis culicivorax*, is a highly promising candidate for use in mosquito control programs. It meets most of the criteria for an excellent biocontrol agent, and it has been utilized effectively in numerous small scale field tests and in a larger test in El Salvador.

New isolations of bacterial pathogens such as *Bacillus thuringiensis* var. *B. sphaericus* show potential for use as larvicides for several species of mosquitoes. There is high commercial interest in *B. thuringiensis* var. *israeliensis*, which is now being produced commercially, and *B. sphaericus* is being produced for laboratory and small scale field tests; thus rapid advances in the development of bacterial pathogens for mosquito control should be expected.

The microsporidian *Nosema algerae* shows high pathogenesis for several *Anopheles* spp. and a few *Culex* spp. Small plot field studies in Panama demonstrated that applications of spores to natural breeding areas of *Anopheles albimanus* resulted in significant infections in natural populations. *Vavraia culicis* infects a wide range of mosquito species, but it has not been studied as thoroughly as *N. algerae*. Other microsporidia in the genera *Amblyospora* and *Parathelohania* have very complex life cycles, and although they are highly host specific and produce high larval mortalities, their biologies are not completely known.

A number of fungal and viral agents have been shown to be pathogenic to mosquito larvae. However, incomplete knowledge to their host specificity, safety, and methods of laboratory production have hampered research.

VIII. REFERENCES

Anderson, J. R. and Ringo, S. L. (1969). *Entomophthora aquatica* sp. n. infecting larvae and pupae of blood-water mosquitoes. *J. Invertebr. Pathol.*, 13, 386-393.

Anthony, D. W., Lotzkar, M. D., and Avery, S. W. (1978). Fecundity and longevity of *Anopheles albimanus* exposed at each larval instar to spores of *Nosema algerae*. *Mosquito News*, 38, 116-121.

Anthony, D. W., Savage, K. E., and Weidhaas, D. E. (1972). Nosematosis: Its effect on *Anopheles albimanus* Wiedemann and a population model of its relation to malaria transmission. *Proc. Helminthol. Soc. Wash.* Special Issue, 39, 428-433.

Anthony, D. W., Savage, K. E., Hazard, E. I., Avery, S. W., Boston, M. D., and Oldacre, S. W. (1978). Field tests with *Nosema algerae* Vavra and Undeen (Microsporida, Nosematidae) against *Anopheles albimanus* in Panama. *Misc. Publ. Entomol. Soc. Am.*, 11, 17-28.

Chapman, H. C. (1974). Biological control of mosquito larvae. *Annu. Rev. Entomol.*, 19, 33-59.

Chapman, H. C. and Glenn, F. E., Jr. (1972) Incidence of the fungus *Coelomomyces punctatus* and *C. dodgei* in larval populations of the mosquito *Anopheles crucians* in two Louisiana ponds. *J. Invertebr. Pathol.*, 19, 256-261.

Chapman, H. C., Clark, T. B., and Petersen, J. J. (1970). Protozoans, nematodes, and viruses of anophelines. *Misc. Publ. Entomol. Soc. Am.*, 7, 134-139.

Clark, T. B. and Chapman, H. C. (1969). A polyhedrosis in *Culex salinarius* of Louisiana. *J. Invertebr. Pathol.*, 13, 312.

Clark, T. B. and Fukuda, T. (1971). Field and laboratory observations of two viral diseases in *Aedes sollicitans* (Walker) in southwestern Louisiana. *Mosquito News*, 31, 193-199.

Clark, T. B., Kellen, W. R., Fukuda, T., and Lindegren, J. E. (1968). Field and laboratory studies on the pathogenicity of the fungus *Beauveria bassiana* to three genera of mosquitoes. *J. Invertebr. Pathol.*, 11, 1-7.

Couch, J. N. (1972). Mass production of *Coelomomyces*, a fungus that kills mosquitoes. *Proc. Nat. Acad. Sci.*, 69, 2043-2047.

Couch, J. N., Romney, S. V., and Rao, B. (1974). A new fungus which attacks mosquitoes and related Diptera. *Mycologia*, 66, 374-379.

Davies, E. E., Howells, R. E., and Venter, D. (1971). Microbial infections associated with plasmodial development in *Anopheles stephensi*. *Ann. Trop. Med. Parasitol.*, 63, 403-408.

de Barjac, H. (1978a). Une nouvelle variété de *Bacillus thuringiensis* trés toxique pour les moustiques: *B. thuringiensis* var. *israelensis* serotype 14. *C. R. Acad. Sci. Paris Ser. D*, 296(10), 797-800.

de Barjac, H. (1978b). Toxicite de *Bacillus thruingiensis* var. *israelensis* pour les larves d'*Aedes aegypti* et d'*Anopheles stephensi*. *C. R. Acad. Sci. Paris Ser. D*, 286(15), 1175-1178.

Federici, B. A. (1973). Virus pathogens of mosquitoes and their potential use in mosquito control. *In* "Mosquito Control" (A. Aubin, J. Bourassa, Bourassa, S. Belloncik, M. Pellissier, and E. Lacoursiere, eds.). *Proc. Int. Seminar on Mosquito Control*, Univ. Quebec, Canada. pp. 93-135.

Federici, B. A. (1975). *Cyclops vernalis* (Copepoda: Cyclopoida) an alternate host for the fungus *Coelomomyces punctatus*. *Proc. Calif. Mosq. Cont. Assoc.*, 43, 172-174.

Federici, B. A. (1977). Virus pathogens of Culicidae (mosquitoes). *In* "Pathogens of Medically Important Arthropods" (D. W. Roberts and M. A. Strand, eds.). *Bull. Wld. Hlth. Org.*, 55 (Suppl. 1), 26-46.

Federici, B. A. and Chapman, H. C. (1977). *Coelomomyces dodgei:* Establishment of an *in vivo* laboratory culture. *J. Invertebr. Pathol.*, 30, 288-297.

Fukuda, T. (1971). *Per os* transmission of *Chilo* iridescent virus to mosquitoes. *J. Invertebr. Pathol.*, 18, 152-153.

Fukuda, T., Lindergren, J. E., and Chapman, H. C. (1976). *Helicosporidium* sp. a new parasite of mosquitoes. *Mosquito News*, 36, 514-517.

Goldberg, L. J. and Margalit, J. (1977). A bacterial spore demonstrating rapid larvicidal activity against *Anopheles sergentii*, *Uranotaenia unguiculata*, *Culex univitattus*, *Aedes aegypti*, and *Culex pipiens*. *Mosquito News*, 37, 355-358.

Hall, D. W. and Anthony, D. W. (1971). Pathology of a mosquito iridescent virus (MIV) infecting *Aedes taeniorhynchus*. *J. Invertebr. Pathol.*, 18, 61-69.

Hall, I. M., Arakawa, K. Y., Dulmage, H. T., and Correa, J. A. (1977). The pathogenicity of strains of *Bacillus thuringiensis* to larvae of *Aedes* and *Culex* Mosquitoes. *Mosquito News*, 37, 246-251.

Hazard, E. I. (1977). Synonomy and host records of Microsporida affecting Culicidae. *In* "Pathogens of Medically Important Arthropods" (D. W. Roberts and M. A. Strand, eds.). *Bull. Wld. Hlth. Org.*, 55 (Suppl. 1), 79-107.

Hazard, E. I. and Chapman, H. C. (1977). Microsporidian pathogens of Culicidae (mosquitoes). *In* Pathogens of Medically Important Arthropods" (D. W. Roberts and M. A. Strand, eds.). *Bull. Wld. Hlth. Org.*, 55 (Suppl. 1), 63-77.

Hazard, E. I. and Fukuda, T. (1974). *Stempellia milleri* sp. n. (Microsporida: Nosematidae) in mosquito *Culex pipiens quinquefasciatus* Say. *J. Protozool.*, 21, 497-504.

Hazard, E. I. and Oldacre, S. W. (1975). Revision of Microsporida (Protozoa) close to *Thelohania* with descriptions of one new family, eight new genera, and thirteen new species. *U.S. Dept. Agric. Tech. Bull.*, 1530, 104 pp.

Hazard, E. I., Andreadis, T. G., Joslyn, D. J., and Ellis, E. A. (1979). Meiosis and its implications in the life cycles of *Amblyospora* and *Parathelohania* (Microspora). *J. Parasitol.*, 65, 117-122.

Ignoffo, C. M., Petersen, J. J., Chapman, H. C. , and Novotny, J. F. (1974). Lack of susceptibility of mice and rats to the mosquito nematode *Reesimermis nielseni* Tsai and Grundmann. *Mosquito News*, 34, 425-428.

Ignoffo, C. M., Biever, K. D., Johnson, W. W., Sanders, H. O., Chapman, H. C., Petersen, J. J., and Woodard, D. B. (1973). Susceptibility of aquatic vertebrates and invertebrates to the infective stage of the mosquito nematode *Reesimermis nielseni*. *Mosquito News*, 33, 599-602.

Kellen, W. R. and Lewallen, L. L. (1969). Response of mosquito larvae to *Bacillus thuringiensis* Berliner. *J. Insect. Pathol.*, 2, 305-307.

Kellen, W. R., Clark, T. B., and Lindegren, J. E. (1963). A possible polyhedrosis in *Culex tarsalis* Coquillett. *J. Insect Pathol.*, 5, 98-103.

Kellen, W. R., Clark, T. B., Lindegren, J. E., and Sanders, R. D. (1966). A cytoplasmic polyhedrosis virus of *Culex tarsalis* (Diptera: Culicidae). *J. Invertebr. Pathol.*, 8, 390-394.

Kellen, W. R., Clark, T. B., Lindegren, J. E., Ho, B. C., Rogoff, M. H., and Singer, S. (1965). *Bacillus sphaericus* Neide as a pathogen of mosquitoes. *J. Invertebr. Pathol.*, 7, 442-448.

Laird, M. (1967). A coral island experiment. A new approach to mosquito control. *Wld. Hlth. Org. Chron.*, 21, 18-26.

Lebedeva, O. P. and Zelenko, A. P. (1972). Virus-like formations in larvae of *Aedes* and *Culex* mosquitoes. *Med. Parazitol. Bolenzi.*, 41, 490-492. (Russian with English summary)

Lebedeva, O. P., Kuznetsova, M. A., Zelenko, A. P., and Gudz-Gorban, A. O. (1973). Investigation of a virus disease of the densonucleosis type in a laboratory culture of *Aedes aegypti*. *Acta Virol.*, 17, 253-256.

Liles, J. N. and Dunn, P. H. (1959). Preliminary laboratory results on the susceptibility of *Aedes aegypti* (Linneaus) to *Bacillus thuringiensis* Berliner. *J. Insect Pathol.*, 1, 309-310.

Linley, J. R. and Nielsen, H. T. (1968). Transmission of mosquito iridescent virus in *Aedes taeniorhynchus*. II. Experiments related to transmission in nature. *J. Invertebr. Pathol.*, 12, 17-24.

McCray, E. M., Jr., Womeldorf, D. J., Husbands, R. C., and Eliason, D. A. (1973). Laboratory observations and field tests with *Lagenidium* against California mosquitoes. *Proc. and Papers of 41st Annu. Conf. Calif. Mosq. Cont. Assoc.*, pp. 123-128.

McNitt, R. E. and Couch, J. N. (1977). *Coelomomyces* pathogens of Culicidae (mosquitoes). *In* "Pathogens of Medically Important Arthropods" (D. W. Roberts and M. A. Strand, eds.), *Bull. Wld. Hlth. Org.*, 55, (Suppl. 1), 123-145.

Petersen, J. J. (1972). Factors affecting sex determination in a mermithid parasite of mosquitoes. *J. Nematol.*, 4, 83-87.

Petersen, J. J. (1973a). Factors affecting mass production of *Reesimermis nielseni* a nematode parasite of mosquitoes. *J. Med. Entomol.*, 10, 75-79.

Petersen, J. J. (1973b). Relationship of density, location of hosts, and water volume to parasitism of larvae of the southern house mosquito by a mermithid nematode. *Mosquito News*, 33, 516-520.

Petersen, J. J. and Chapman, H. C. (1970). Parasitism of *Anopheles* mosquitoes by a *Gastromermis* sp. (Nematoda: Mermithidae) in southwestern Louisiana. *Mosquito News*, 30, 420-424.

Petersen, J. J. and Willis, O. R. (1972). Results of preliminary field applications of *Reesimermis nielseni* to control mosquito larvae. *Mosquito News*, 32, 312-316.

Petersen, J. J. and Willis, O. R. (1974a). Experimental release of a mermithid nematode to control *Anopheles* mosquitoes in Louisiana. *Mosquito News*, 34, 316-319.

Petersen, J. J. and Willis, O. R. (1974b). *Diximermis peterseni* (Nematoda: Mermithidae): A potential biocontrol agent of *Anopheles* mosquito larvae. *J. Invertebr. Pathol.*, 24, 20-23.

Peterson, J. J., Willis, O. R., and Chapman, H. C. (1978). Release of *Romanomermis culicivorax* for the control of *Anopheles albimanus* in El Salvador. I. Mass production of the nematode. *Am. J. Trop. Med. Hyg.*, 27, 1265-1267.

Petersen, J. J., Chapman, H. C., Willis, O. R., and Fukuda, T. (1978). Release of *Romanomermis culicivorax* for the control of *Anopheles albimanus* in El Salvador. II. Application of the nematode. *Am. J. Trop. Med. Hyg.*, 27, 1268-1273.

Ramoska, W. A., Burgess, J., and Singer, S. (1978). Field application of bacterial insecticide. *Mosquito News*, 38, 57-60.

Roberts, D. W. (1974). Fungal infections in mosquitoes. *In* "Mosquito Control" (A. Aubin, J. Bourassa, S. Belloncik, M. Pellissier, and E. Lacoursiere, eds.). *Proc. Int. Seminar on Mosquito Control*, Univ. Quebec, Canada. pp. 143-193.

Roberts, D. W. (1975). Isolation and development of fungus pathogens of vectors. *In* "Biological Regulation of Vectors" (J. D. Briggs, ed.). *Dept. HEW Publication* No. (NIH)77-1180. pp. 85-93.

Roberts, D. W. (1977). Fungal pathogens, except *Coelomomyces*, of Culicidae (mosquitoes). *In* "Pathogens of Medically Important Arthropods" (D. W. Roberts and M. A. Strand, eds.). *Bull. Wld. Hlth. Org.*, 55 (Suppl. 1), 147-172.

Singer, S. (1973). Insecticidal activity of recent bacterial isolates and their toxins against mosquito larvae. *Nature (London)*, 244, 110-111.

Singer, S. (1975). Isolation and development of bacterial pathogens of vectors. *In Proc. Natl. Inst. Hlth Wkshp*, October 1975.

Singer, S. (1977). Bacterial pathogens of Culicidae (mosquitoes). *In* "Pathogens of Medically Important Arthropods" (D. W. Roberts and M. A. Strand, eds.). *Bull. Wld. Hlth. Org.*, 55 (Suppl. 1), 47-62.

Singer, S. and Murphy, L. J. (1976). New insecticidal strains of *Bacillus sphaericus* useful against *Anopheles albimanus* larvae. *Proc. Annu. Mtg. Am. Soc. Microbial.*, Atlantic City.

Steelman, C. D., White, T. W., and Schilling, P. E. (1972). Effects of mosquitoes on the average daily gain of feedlot steers in southern Louisiana. *J. Econ. Entomol.*, 65, 462-466.

Steelman, C. D., White, T. W., and Schilling, P. E. (1973). Effects of mosquitoes on the average daily gain of Hereford and Brahman bred steers in southern Louisiana. *J. Econ. Entomol.*, 66, 1081-1083.

Stoltz, D. B., Fukuda, T., and Chapman, H. C. (1974). Virus-like particles inthe mosquito *Culex salinarius*. *J. Microsc.*, 19, 109-112.

Umphlett, C. J. (1970). Infection levels of *Coelomomyces punctatus*, an aquatic fungus parasite, in a natural population of the common malaria mosquito, *Anopheles quadrimaculatus*. *J. Invertebr. Pathol.*, 15, 299-305.

Weiser, J. and Coluzzi, M. (1972). The microsporidian *Plistophora culicis* Weiser 1946 in different mosquito hosts. *Folia Parasitol.*, 19, 197-202.

Whisler, H. C., Zebold, S. L., and Shemanchuck, J. A. (1974). Alternate host for mosquito parasite *Coelomomyces*. *Nature*, 251, 715-716.

Whisler, H. C., Zebold, S. L., and Schemanchuck, J. A. (1975). Life history of *Coelomomyces psorophorae*. *Proc. Nat. Acad. Sci.*, 72, 963-966.

Woodard, D. B. (1978). Establishment of the nematode *Diximermis peterseni* in the field in southwest Louisiana using laboratory reared material. *Mosquito News*, 38, 80-83.

SOME PROTOZOA INFECTING FIRE ANTS, *Solenopsis* SPP

Donald P. Jouvenaz

Insects Affecting Man and Animals Research Laboratory
Agricultural Research
Science and Education Administration
U.S. Department of Agriculture
Gainesville, Florida

I. INTRODUCTION

The black and the red imported fire ants, *Solenopsis richteri* and *Solenopsis invicta*, are medically and agriculturally important insects that infest ca. 8 x 10^7 ha (2 x 10^8 acres) in the southeastern United States. Both species were introduced into the United States from South America at Mobile, Alabama; the black imported fire ant about 1918 and the red imported fire ant about 1940. *Solenopsis richteri* is now restricted to a relatively small area in northeastern Mississippi and northwestern Alabama; *S. invicta* infests Florida and Louisiana, and parts of North Carolina, South Carolina, Georgia, Alabama, Mississippi, Texas, and Arkansas. The tropical fire ant, *Solenopsis geminata*, may also be an introduced species; however, it has been a resident of this country for so long it is generally regarded as native.

The southern and desert fire ants, *Solenopsis xyloni* and *Solenopsis aurea*, are truly native to the United States. None of these latter three species are important as pests, except in Hawaii where *S. geminata* has been introduced.

II. RECENT EFFORTS TO IDENTIFY POTENTIAL PATHOGENS OF FIRE ANTS

Efforts to control the imported fire ants by chemical means have been the subject of serious controversy for more than 15 years (Lofgren et al., 1975). Consequently, surveys for pathogens of *S. richteri* were conducted in Mississippi (Broome, 1974) and of *S. invicta* in Florida (B. A. Federici, pers. commun.). These surveys failed to identify pathogens other than ubiquitous, non-specific, facultative organisms such as the bacterium *Serratia marcescens*. Recently, Jouvenaz et al. (1977) made an extensive survey in six states, the results of which confirmed the rarity of disease in these ants in the United States.

In a sample of 1007 *S. invicta* colonies from 285 collection sites in six states, only one colony was infected with a bona fide pathogen, a microsporidium. Since this parasite was found infecting four colonies of *S. gaminata* (in a sample of 307) from collection sites in three Florida counties, it seems probable that *S. geminata* is the normal host. (We have transmitted one parasite of *S. geminata*, the microsporidium *Burenella dimorpha*, to *S. invicta* in the laboratory; however, the infection does not persist in *S. invicta* colonies. Thus, the naturally occurring infection also may have been abortive.) Since the conclusion of the survey, several hundred additional colonies of *S. invicta* (primarily from central Florida) have been examined; all were devoid of pathogens.

Similarly, no potential pathogens were found in 83 colonies of *S. richteri*. (Although this sample was small it actually represents a more intesive sampling of this species because *S. richteri* is restricted to a very small area in northeastern Mississippi and northwestern Alabama.)

An unidentified endozoic yeast was associated with 93 (9.25%) of the colonies of *S. invicta*. These yeast cells occur free in the hemolymph of immatures and adults. With proper care, laboratory colonies harboring this organism can live quite well; however, with neglect or stress they appear to have elevated mortality rates. The yeast grows poorly on standard mycological media, but grows well in insect tissue culture media. Yeasts grown *in vitro* can be transmitted *per os* to healthy colonies of *S. invicta*. Taxonomic studies of the yeast and an evaluation of its pathogenicity and host range are in progress at this laboratory. (Endozoic yeasts have not yet been detected in fire ants in South America.)

Virus-like particles have been detected in *S. geminata* and in an undescribed *Solenopsis* sp. from Brazil (Avery et al., 1977). The pathogenicity of these viruses is as yet undetermined. Ultimately, we seek to establish a complex of natural enemies (including enemies othern than pathogens) that will exert continuing stress or control of the fire ants. To be of value, pathogens and other natural enemies of fire ants need not be rapidly fatal to colonies or to large numbers of individual ants. Debilitating diseases may enable our native ants which compete with fire ants to do so more successfully. Surveys for natural enemies and extensive ecological studies over a period of several years in South America are planned.

The role of pathogens in fire ant population dynamics in South America is unknown. Indeed, little is known even about fire ant population densities in various habitats.

The incidence of disease among colonies in fire ants in South America may be limited, at least in part, by the dry season, which presumably is a period of stress. Reasonably, then, one would expect higher mortality among diseased colonies than among healthy colonies during this time. Thus, the overall incidence of disease may be lowest at the end of the dry season. During the wet season, the environment presumably is much more favorable so the numbers of colonies and, after a lag period, the incidence of disease, may increase. Recurrence of the dry season would prevent disease from reaching epizootic proportions.

In the United States, extended dry seasons comparable to those of Mato Grosso do not occur, though our winters may exert similar control of epizootics. Nevertheless, severe, extended winters do not occur in much of the area infested by fire ants in the United States. Other (unknown) abiotic or biotic factors that limit the incidence of desease in South America may also not be operative in this country. We plant to conduct studies of population dynamics of fire ants and their pathogens and natural enemies in South America. However, we are very cautious in extrapolating the findings to our different environment.

The imported fire ants in the United States present the classic situation for imported pests that have been freed of natural enemies. The identification, importation, and establishment of a complex of natural enemies may ameliorate this problems.

III. PROTOZOAN PATHOGENS

The first observation of a protozoan infection in fire ants was made by W. F. Buren during a taxonomic study of *S. invicta* from Mato Grosso, Brazil (Allen and Buren, 1974). While he was examining alcohol-preserved specimens, Buren observed subspherical, cyst-like bodies in the partially cleared gasters of worker ants. These bodies were found to contain spores of a microsporidium that was subsequently named *Thelohania solenopsae* by Knell et al. (1977) who described the parasite from fresh material. Soon after this initial observation, Allen and Silveira-Guido (1974) reported similar microsporidia infecting *S. richteri* in Uraguay and Argentina and an unidentified *Solenopsis* sp. in Uraguay.

T. solenopsae infects fat body cells of workers and sexuals and the ovaries of queens. Infected cells hypertrophy to form the cysts observed by (Allen and Buren, 1974). The disease is not rapidly fatal, but the destruction of the fat body that occurs results in premature death of adult ants. Consequently, colonies are debilitated.

T. solenopsae is the only member of its genus that is dimorphic (Knell et al., 1977). Two types of spores develop simultaneously in the same tissues: uninucleate, membrane-bounded (MB) spores in octets, and binucleate, nonmembrane-bounded (NMB) spores. Of the remaining 7 genera of the family Thelohaniidae, at least two, *Parathelohania* and *Amblyospora*, are dimorphic; however, their NMB spores are uninucleate (Hazard and Oldacre, 1974).

Atempts to transmit *T. solenopsae per os* failed (unpubl.).

T. solenopsae has been detected in 22 described and undescribed species of fire ants in South America (Jouvenaz et al., 1977). In many areas, it infects 25% or more of the colonies; other pathogens, including protozoa, also occur in these ants. However, there are no data concerning the effects of diseases on fire ant populations in South America.

Four species of microsporidia and one neogregarine are known to infect *S. geminata* in the southeastern United States (Jouvenaz et al., 1977). One microsporidium, *Burenella dimorpha*, was recently described by Jouvenaz and Hazard (1978) was the type species of a new genus that represent a new family (Burenellidae). This family is characterized by two sequences of sporogony that produce morphologically different sporonts and spores. One sequence has disporous sporonts that sporulate in the hypodermis, producing binucleate NMB spores. The second sequence has multinucleate sporonts that produce 8 uninucleate MB spores in fat cells. The species of Burenellidae are distinguished from the dimorphic species of Thelohaniidae by the mode of formation of NMB spores. The NMB spores of Burenellidae arise from disporous sporonts, while in Thelonahiidae a variable number (6-40) of NMB spores arise from plasmodia. Also, the MB sporonts of species of Burenellidae do not secrete granules in the lumen of the pansporoblast as do those of Thelohaniidae. In *B. dimorpha*, the pansporoblastic membrane enclosing the spores in the second sequence is fragile and sub-persistent, and consequently MB spores are not seen in octets by light microscopy. NMB spores begin development earlier in the course of infection than do MB spores and always predominate in number. Typically, they consitute ca. 60-70% of the total spores in advanced infections, but occasionally may constitute more than 99% of the spores in individual ants.

The cycle of *B. dimorpha* infection within an ant colony is as follows: NMB spores develop in the hypodermis, which is destroyed, producing clear areas in the heads, petioles, and gasters of pupae (Fig. 1). As the infection progresses, the cuticle becomes very fragile and eventually ruptures. The adult ants cannibalize these ruptured pupae but do not ingest the spores. Instead, the spores, together with other particulate matter, are diverted to the infra-buccal cavity and formed into an infrabuccal pellet. This pellet is expelled and placed on a specialized anteroventral area, the praesaepium of fourth-instar larvae. The praesaepium, which bears spines specialized for holding solid food while the larva feeds, is not present on earlier instars, which feed only on liquids. Because of this method of feeding, the fourth-instar larva is the only stage which is vulnerable to infection. Both spore types are ingested, but only the NPMB spore is infective. The PMB spores are expelled unextruded in the meconium upon pupation.

The function of the PMB spore remains unknown. A most attractive hypothesis is that it either infects an alternate host or is primed in the gut of a mechanical vector for extrusion upon subsequent ingestion by ant larvae. Either would explain the mode of intercolonial transmission of the infection (fire ants are territorial and aggressive toward conspecific ants) (Jouvenaz et al., 1981). Many candidate species exist for the role of vector; a large and varied arthropod fauna is associated with fire ants. Collins and Markin (1971) listed 52 species of insects that have been collected from fire ant nests; other invertebrates also occur. At least some of these organisms have symbiotic relation-ships with fire ants and are known to travel between fire ant nests (Wojcik, 1975).

B. dimorpha has been transmitted *per os* to *S. xyloni*, *S. invicta*, and *S. richteri* as well as to its normal host, *S. geminata*. All four species are readily infected by feeding them boiled egg yolk wetted with suspension of spores; however, the infection does not persist in colonies of *S. xyloni*, *S. invicta*, or *S. richteri*. Apparently the cuticle does not rupture as readily in these species, and the infection does not spread as

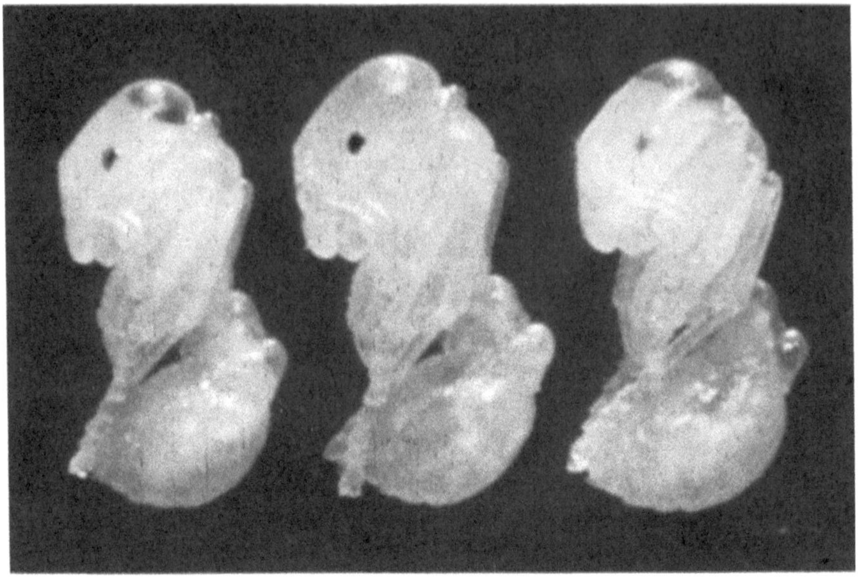

Fig. 1. Pathologic changes in pupae (left and right) infected with *Burenella dimorpha*. Note the clear areas in the occipital region of the head and petiole. The pupa in the center is uninfected.

efficiently to larvae. In other respects, the course of infection appears similar in all four species of ants.

The remaining microsporidia of *S. geminata* infect oenocytes and/or fat cells, causing hypertrophy but not formation of cysts as occurs in *T. solenopsae* infection. At least one of the three undescribed species of microsporidia of *S. geminata* is dimorphic, with the spore types developing in different fat body cells. Ants infected by these microsporidia survive to the adult stage. Similar microsporidia occur in fire ants in South America.

The neogregarine of *S. geminata* is a new species of *Mattesia* (Neogregarinida: Ophrocystidae). It develops in the oenocytes of the hypodermis, causing destruction of the hypodermis, melanization, and eye malformation of pupae (Jouvenaz and Anthony, 1979). The disease appears to be invariably fatal in the pupal stage of development. Attempts to transmit the infection *per os* (using fresh, aged, and variously treated spores) or by placing infected pupae in healthy colonies (conspecific pupae are adopted) have failed thus far. A similar or identical neogregarine also occurs in fire ants from South America.

The protozoan pathogens of *S. geminata* do not appear to offer promise as biological control agents for the imported fire ants, since *S. invicta* and *S. richteri* have been residents of the United States for ca. 40 and 60 years, respectively, and have not permanently acquired any of the diseases of *S. geminata*. Experimental laboratory and field transmission, where successful, have not produced infections that persisted and spread in the *S. invicta* population. Adaptation of a parasite of *S. geminata* to *S. invicta* through strain selection may, of course, be possible.

Our primary purpose in studying the diseases of *S. geminata* is to gain basic knowledge of the taxonomy and biology of fire ant pathogens, and to develop techniques for the propagation, study, and possible eventual introduction of exotic pathogens, especially those of *S. invicta* and the *S. invicta* complex.

IV. CONCLUSIONS

The red and black imported fire ants, *S. invicta* and *S. richteri*, are medical and agricultural pests that infest ca. 8×10^7 ha in nine southeastern states. In their native South America, these ants are parasitized by various protozoa and other pathogens; in the United States they appear to be essentially free of natural enemies. Thus, they present the classic situation with regard to imported pests, and the problem may be ameliorated by the introduction of a complex of natural enemies.

In South America, the most common pathogen of fire ants is *Thelohania solenopsae*. This microsporidium infects 20-25% or more of the colonies in many areas. The sites of infection are in the fat body cells of workers and sexuals and in the ovaries of the queens. The infected cells hypertrophy and form cysts. The disease is not rapidly fatal, but destruction of the fat body occurs, which results in premature dealth of adult ants. Consequently, colonies are debilitated.

The tropical fire ant, *S. geminata*, a species native to the United States, is the host of our microsporidia, of a neogregarine, and of a virus. These organisms lack potential as biological control agents for imported fire ants, but they are being studied as models.

V. REFERENCES

Allen, G. E. and Buren, W. F. (1974). Microsporidan and fungal diseases of *Solenopsis invicta* Buren in Brazil. *J. N. Y. Entomol. Soc.*, 82, 125-130.

Allen, G. E. and Silveira-Guido, A. (1974). Occurrence of microsporida in *Solenopsis richteri* and *Solenopsis* sp. in Uruguay and Argentina. *Fla. Entomol.*, 57, 327-329.

Avery, S. W., Jouvenaz, D. P., Banks, W. A., and Anthony, D. W. (1977). Virus-like particles in a fire ant, *Solenopsis* sp. (Hymenoptera: Formicidae) from Brazil. *Fla. Entomol.*, 60, 17-20.

Broome, J. R. (1974). Microbial control of the imported fire ant, *Solenopsis richteri* Forel. *Ph.D. Dissert.* Mississippi State University, Mississippi State, Mississippi. Dissert. Abst. Int. B35, 3954.

Collins, H. L. and Markin, G. P. (1971). Inquilines and other arthropods collected from nests of the imported fire ant, *Solenopsis saevissima richteri*. *Ann. Entomol. Soc. Amer.*, 64, 1376-1380.

Hazard, E. I. and Oldacre, S. W. (1974). Revision of microsporida (Protozoa) close to *Thelohania*, with descriptions of one new family, eight new genera, and thirteen new species. *USDA Tech. Bull.*, 1530, 104 pp.

Jouvenaz, D. P., Allen, G. E., Banks, W. A., and Wojcik, D. P. (1977). A survey for pathogens of fire ants, *Solenopsis* spp., in the Southeastern United States. *Fla. Entomol.*, 60, 275-279.

Jouvenaz, D. P. and Hazard, E. I. (1978). New family, genus, and species of microsporida (Protozoa: Microsporida) from the tropical fire ant, *Solenopsis geminata* (Fabricius) (Insecta: Formicidae). *J. Protozool.*, 25, 24-29.

Jouvenaz, D. P. and Anthony, D. W. (1979). *Mattesia geminata* sp. n. (Neogregarinida: Ophrocystidae) a parasite of the tropical fire ant, *Solenopsis geminata* (Fabricius). *J. Protozool.*, 26, 354-356.

Jouvenaz, D. P., Lofgren, C. S., and Allen, G. E. (1981). Transmission and infectivity of spores of *Burenella dimorpha* (Microsporida: Burenellidae). *J. Invertebr. Pathol.*, 37, 265-268.

Knell, J. D., Allen, G. E., and Hazard, E. I. (1977). Light and electron microscope study of *Thelohania solenopsae* sp. n. in the red imported fire ant, *Solenopsis invicta* Buren. *J. Invertebr. Pathol.*, 29, 192-200.

Lofgren, C. S., Banks, W. A., and Glancye, B. M. (1975). Biology and control of imported fire ants. *Ann. Rev. Entomol.*, 20, 1-30.

Wojcik, D. P. (1975). Biology of *Myrmecaphodius excavaticollis* (Blanchard) and *Euparia castanea* Serville (Coleoptera: Scarabaeidae) and their relationships to *Solenopsis* spp. (Hymenoptera: Formicidae). *Ph.D. dissertation,* University of Florida, Gainesville. *(Dissert. Abst. Int. B.,* 36, 5962).

VERTICAL TRANSMISSION OF PATHOGENS OF INVERTEBRATES

Paul E.M. Fine

Ross Institute
London School of Hygiene and Tropical Medicine
London, England

I. INTRODUCTION

It has long been recognized that many pathogens of invertebrates can be transmitted directly from infected hosts to their immediate progeny. Indeed, this subject has achieved considerable eminence in the literature pertaining to infectious diseases, having been discussed at length by Louis Pasteur (1870) in his classical work on the diseases of silkworms entitled *Etudes sur la Maladie des Vers a Soie*. In this work, which deserves as much credit as any document to be considered a cornerstone of the germ theory, Pasteur recognized that the transfer of the pébrine agent - the microsporidian now known as *Nosema bombycis* - in silkworm eggs provided the key to the distribution and to the practical resolution of an important agricultural problem. Table 1 represents a translation of one of Pasteur's tables showing data on infection rates in samples of silkworm eggs sent to France from Japan. This type of information allowed Pasteur to outline a method for routine sampling of egg batches in order to prevent the introduction of pebrine into silkworm colonies. Since Pasteur's time many workers have noted that propensity for hereditary transmission of insect pathogens, and this phenomenon has accumulated a sizable literature. In addition to its role in diseases of beneficial insects, hereditary transmission is recognized to be important in the spatial and temporal distribution of many natural infections of noxious insects, and hence of potential importance in biological control.

Most of the literature on this subject is descriptive, rather than analytical, emphasizing descriptions of agents and their life cycles in a variety of invertebrate hosts. The propensity for transmission from parent to progeny is often noted merely as a curiosity, although its common occurrence has led many workers to search specifically for this feature when describing a new infectious agent. Such studies have revealed that different agents may employ different mechanisms in achieving transfer from parent to progeny - some transferring directly from the female parent inside the egg (now generally called transovarial transmission), some transferring directly from the female parent on the exterior of the egg (called transovum transmission), and a few transferring directly from the male parent via sperm or seminal fluid or else indirectly via venereal infection of the female (the transovarial-transovum distinctions are taken from Steinhaus and Martignoni, 1970). In some situations these differences in mechanisms may be important. For example, transovum transmission may be preventable by surface sterilization of eggs, whereas transovarial transmission is not. In many circumstances, however, it is not the differences, but the similarities between such transmission mechanisms which are important, as each ensures transfer of an agent from a parent to offspring. Emphasis on this similarity has led to increasing use of the term "vertical transmission" to cover all mechanisms for the direct partent-to-offspring transfer of infective agents.

The term "vertical transmission", which is now widely used in the biological literature, was first used in the context of cancer studies (Gross, 1944). It is essentially an epidemiological concept. Its reference is to a pattern of transfer of infection within a host population, the "vertical" referring to the direction of parent-offspring connections in a conventional pedigree diagram as indicated in Figure 1. Other terms also are used to describe

TABLE 1. Extract from one of Pasteur's tables showing results of examining silkworm eggs for *Nosema bombycis* "Corpuscles". These eggs were sampled from lots imported into France from Japan.[a]

Number Inscribed on Carton	EGGS OF BAD APPEARANCE		EGGS OF GOOD APPEARANCE	
	No. of eggs examined	No. of eggs with "corpuscules"	No. of eggs examined	No. of eggs with "corpuscules"
A1 161	3	1	33	NONE
A1 162	3	1	18	NONE
A1 164	3	NONE	33	1
A2 564	13	NONE	33	NONE

[a]Original in French on page 477 of the 1926 edition of Pasteur (1870).

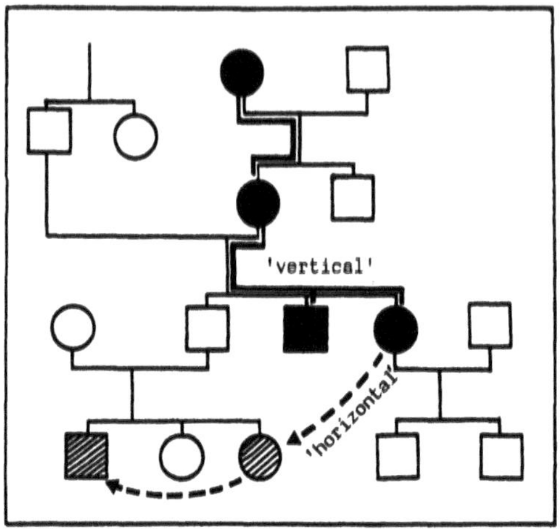

Figure 1. Conventional pedigree diagram showing four successive generations of a host species. Solid circles (females) and squares (males) are infected with an infectious agent by vertical transmission (solid arrows). Hatched symbols are infected by horizontal transmission (broken arrows).

this phenomenon. For example, "generation-to-generation transmission" is widely used in the invertebrate pathology literature. It may be pointed out that insofar as the phrase "generation-to-generation" need not necessarily imply a parent-to-offspring relationship (e.g., an aunt and her niece are often considered to belong to successive generations), the term vertical transmission, defined as direct transfer of infection from a parent organism to his or her progeny, is preferable. The definition relegates all transmission between individuals that are not parent-to-offspring relationships whether within or between generations - to the category of non-vertical ("horizontal") transmission.

This review concentrates on the epidemiological[1] implications of the vertical transmission of infections of invertebrates. Most of the agents discussed are well recognized pathogens of invertebrates. The paper does not emphasize the vertical transfer of animal or plant pathogens in their arthropod vectors since this subject has been dealt with at length elsewhere (Burgdorfer and Varma, 1967; Fine, 1975, 1979a). On the other hand, several infections of uncertain or questionable pathogenicity are included to illustrate the variety of different situations resulting from mutualism, commensalism, or parasitism. Moreover, the inclusion of non-pathogens allows for a discussion of the very important problem of defining and measuring pathogenicity in general, a question which will be found to play a crucial role in assessing the implications of vertical transmission phenomena.

II. EXTENT OF THE PHENOMENON: SURVEY AND CLASSIFICATION

A review of the literature on the vertical transmission of invertebrate pathogens is a tantalizing but disappointing exercise: tantalizing in its disclosure of a large number and variety of examples of this phenomenon, but disappointing in the paucity of useful information available. Most of the relevant literature consists of observations which suggest that vertical transmission occurs with one or another infection and under varying conditions. Some literature is available on anatomical pathways of infection in terms of distinguishing intra- versus extra-ovum transmission of

[1]The author maintains strong preference for the term epidemiological, rather than "epizootiological". As the basic science concerned with the study of the distribution and determinants of diseases in populations, "epidemiology" is applicable to all species - and is indeed generally used in the veterinary (Schwabe et al., 1977) and plant disease (Van der Plank, 1963) fields, in addition to the human medical field. The use of "epizootiology" by certain groups of workers has had the unfortunate result of isolating them from the fundamental concepts, methodology, and literature appropriate for the study of disease in populations. The misuse and confusion of incidence and prevalence measures, throughout the invertebrate pathology literature, is a clear example of this unhappy situation. In addition, "epidemiology" is more easily pronounced!

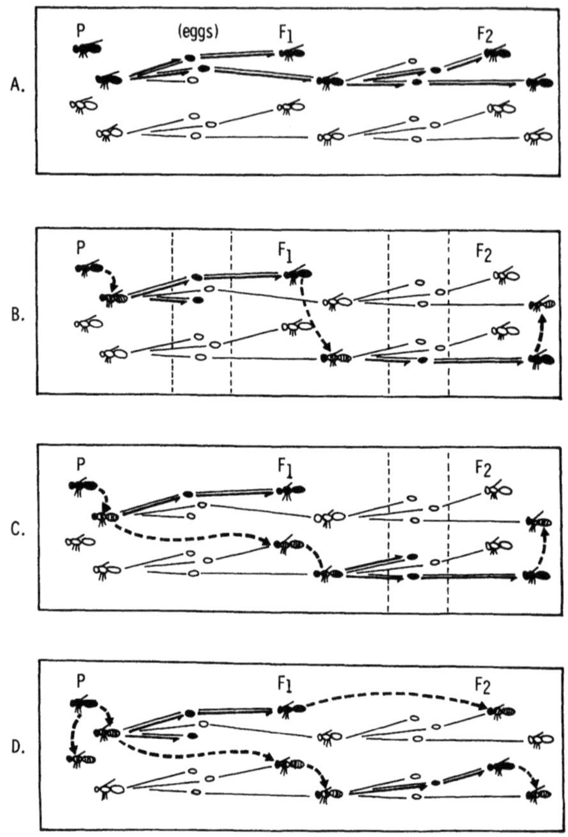

Figure 2. Patterns of infection transmission within and between three successive generations (P, F_1, F_2) of an insect host species. Solid cartoon figures are infected by vertical transmission (solid arrows) and hatched cartoon figures are infected by horizontal transmission (broken arrows). A) Type A infection transmitted by vertical pathway only; B) Type B vertical transmission between generations and horizontal transmission within generations; C) Type C vertical transmission between generations and horizontal transmission within all and between some generations; D) Type D vertical transmission between generations and horizontal transmission within and between all generations.

the agents. But there is (with a few notable exceptions) very little quantitative information on transmission rates or analytical discussion on the extent of the phenomenon in nature or in the laboratory. The published comments on the epidemiological implications of such transmission have been, in general, speculative.

Several examples of vertically transmitted pathogens of invertebrates are cited in Table 2. For the purposes of this review, these examples have been classifed into four groups, designated as transmission types A, B, C, and D, that depend on the qualitative role played by the vertical transmission pathway in maintaining the infectious agents in nature. Types A, B, and D correspond to transmission types II-a, II-b, and II-c as described previously (Fine, 1975).

The four transmission types are illustrated diagramatically in Figure 2. The diagrams present various combinations of pathways in which an infectious agent may be transmitted from one invertebrate host to another host. Since the main subject of this paper deals with vertical transfer (i.e., directly from parent to progeny), the diagrams illustrate groups of parental-generation host insects (to the left) and groups of their eggs and immediate progeny (to the right). (It may be noted that horizontal transfer of invertebrate pathogens often involves the ingestion of agents released into the environment by infected or deceased individuals).

In the type A transmission pattern, vertical transmission of an infectious agent is the only means of transfer between hosts. Thus infection is strictly the result of hereditary involvement and is exemplified by the many mutualistic bacteroid symbiotes of insects and other invertebrates. These symbiotes have been extensively reviewed by Buchner (1965). Most of the agents classified in this group are thought to be transmitted only by the female parent; that is, their inheritance is strictly "matroclinal". It has been shown that strict maternal inheritance is responsible for maintaining agents which impart a selective advantage to (i.e., are "mutuals" in) their host organisms (Fine, 1974, 1975). A few hereditary infections have evolved mechanisms which allow them to escape this requirement for direct selective benefit to the host, either by undergoing transfer from male as well as the female parent (e.g., sigma virus of *Drosophila*), or by providing indirect frequency-dependent advantages to their hosts (e.g., *Wolbachia* in culicine mosquitoes, or sex ratio spirochaetes in *Drosophila*). These important exceptions are discussed later.

It should be noted that some authors have claimed that certain microsporidia, particularly certain members of the genus *Thelohania*, may be maintained by vertical transmission alone and may be

TABLE 2. Vertically transmitted infections of invertebrates classified according to the role of the vertical transmission pathway in maintaining the infectious agent within the host population.[a]

TYPE A TRANSMISSION: Infection transmitted only by vertical transmission.

Agent	Host	Inside (i) or outside (o) egg	References
Sigma Virus	*Drosophila melanogaster*	i	l'Heritier, P. (1970). Seecof, R. L. (1968).
Mycoplasma-like organisms (MLO)	*Drosophila paulistorum*	i	Ehrman, L. and Kernaghan, R. P. (1971).
Sex Ratio (SR) (cytoplasmic)	*Drosophila bifasciata*	i	Poulson, D. F. (1963).
Sex Ratio (SR) (spirochaete)	*Drosophila paulistorum*	i	Poulson, D. F. (1963).
	Drosophila willistoni	i	Malogolowkin, G. and Poulson, D. F. (1958). Poulson, D. F. (1963). Sakaguchi, B. and Poulson, D. F. (1963).
Wolbachia pipientis	*Culex pipiens*	i	Fine, P. E. M. (1978a). Singh, K. R. P., Curtis, C. F., and Krishnamurthy, B. S. (1976).

Agent	Host	Inside (i) or outside (o) egg	References
Mutual bacteroids	Many species		Buchner, P. (1965).
TYPE B TRANSMISSION:	Vertical transmission between generations and horizontal transmission within generations.[b]		
Keystone virus	*Aedes atlanticus*	i	Fine, P. E. M. and LeDuc, J. W. (1978).
TYPE C TRANSMISSION:	Vertical transmission between generations and horizontal transmission within all and between some (but not all) generations.		
Mosquito iridescent viruses	*Aedes spp.*	i	Chapman, H. C., Clark, T. B., Petersen, J. J., and Woodard, D. B. (1969). Hall, D. W. and Anthony, D. W. (1971). Linley, J. R. and Nielsen, H. T. (1968a). Linley, J. R. and Nielsen, H. T. (1968b). Woodard, D. B. and Chapman, H. C. (1968).
Polyhedrosis virus	*Neodiprion sertifer* (?)		Bird, F. T. (1961).
Perezia fumiferanae	*Choristoneura fumiferana*		Thompson, H. M. (1958).

(continued)

TABLE 2. (Continued)

Agent	Host	Inside (i) or outside (o) egg	References
Nosema bombycis	*Bombyx mori*		Pasteur, L. (1870).
TYPE D TRANSMISSION:	Vertical transmission between generations and horizontal transmission within and between all generations.		
VIRUSES			
Nuclear Polyhedrosis virus	*Aedes sollicitans*		Clark, T. B. and Fukuda, T. (1971).
	Colias eurytheme	o	Martignoni, M. E. and Milstead, J. E. (1962).
	Diprion hercyniae		Bird, F. T. (1961). Nielson, M. M. and Elgee, D. E. (1968).
	Heliothis zea	o	Hamm, J. J. and Young, J. R. (1974).
	Lymantria dispar	o	Doane, C. C. (1969). Doane, C. C. (1970a). Doane, C. C. (1970b). Doane, C. C. (1975).
	Neodiprion lecontei		Bird, F. T. (1961).

Agent	Host	Inside (i) or outside (o) egg	References
Nuclear Polyhedrosis virus	*Neodiprion swainei*	o	Smirnoff, W. A. (1961). Smirnoff, W. A. (1962).
	Orgyia leucostigma	i?	Bird, F. T. (1961).
	Prodentia litura	o	Harpaz, I. and Ben Shaked, Y. (1964).
	Spodoptera exempta	o	Brown, E. S. and Swain, G. (1965). Swaine, G. (1966).
	Trichoplusia ni	o	Vail, P. V. and Hall, I. M. (1969).
Cytoplasmic polyhedrosis virus	*Aedes sollicitans*		Clark, T. B. and Fukuda, T. (1971).
	Bombyx mori	i	Aruga, H. and Nagashima, E. (1962). Hukuhara, T. (1962).
	Heliothis virescens	o	Sikorowski, P. P., Andrewes, G. L., and Broome, J. R. (1973).
	Pectinophora gossypiella	o	Bullock, H. R., Mangum, C. L., and Guerra, A. A. (1969).
	Trichoplusia ni	o	Vail, P. V. and Gough, D. (1970).

(continued)

TABLE 2. (Continued)

Agent	Host	Inside (i) or outside (o) egg	References
Granulosis virus	*Laspeyresia pomonella*		Etzel, L. K. and Falcon, L. A. (1976).
	Pieris brassicae	o	David, W. A. L., Gardiner, B. O. C., and Clothier, S. E. (1968). David, W. A. L. and Taylor, C. E. (1976).
Nonoccluded virus	*Culex tarsalis*		Richardson, J., Sylvester, E. S., Reeves, W. C., and Hardy, J. L. (1974).
	Teleogryllus spp.	o	Reinganum, C., O'Loughlin, G. T., and Hogan, T. W. (1970).
Virus	*Danaus plixippus*		Urquhart, F. A. (1966).
MICROSPORIDIA			
Amblyospora sp.	*Culex salinarius*	i	Andreadis, T. G. and Hall, D. W. (1979a). Andreadis, T. G. and Hall, D. W. (1979b).
Nosema spp.	*Anopheles stephensi*	o	Alger, N. E. and Undeen, A. H. (1970). Canning, E. U. and Hulls, R. H. (1970).
	Estigmene acrea		Nordin, G. L. and Maddox, J. V. (1974).

Agent	Host	Inside (i) or outside (o) egg	References
Nosema spp.	*Heliothis zea*	i	Brooks, W. M. (1968). Gaugler, R. R. and Brooks, W. M. (1975).
	Ornithodorus parkeri (*tick*)		Krinsky, W. L. (1977).
	Plodia interpunctella	i	Kellen, W. R. and Lindegren, J. E. (1971). Kellen, W. R. and Lindegren, J. E. (1973).
	Schizocerrella pilicornis	i	Gorske, S. F. and Maddox, J. V. (1978).
Perezia pyraustae	*Pyrausta nubialis*		Zinmmack, H. L., Arbuthnot, K. D., and Brindley, T. A. (1954). Zimmack, H. L. and Brindley, T. A. (1957).
Pleistophora spp.	*Armanda brevis* (*polychaete*)		Szollosi, D. (1971).
	Culiseta inornata		Chapman, H. C. and Kellen, W. R. (1967).
Stempellia spp.	*Culex restuans*		Anderson, J. F. (1968).
Thelohania spp.	*Aedes cantator*		Anderson, J. F. (1968).
	Chaoborus astictopus		Sikorowski, P. P. and Madison, C. H. (1968).

(continued)

TABLE 2. (Continued)

Agent	Host	Inside (i) or outside (o) egg	References
	Culex tarsalis	i	Kellen, W. R. (1962). Kellen, W. R. and Wills, W. (1962).
Various microsporidia	Various mosquitoes		Chapman, H. C. (1974). Chapman, H. C., Woodard, D. B., Kellen, W. R., and Clark, T. B. (1966). Chapman, H. C., Woodard, D. B., and Petersen, J. J. (1967). Fox, R. M. and Weiser, J. (1959). Hazard, E. I. and Weiser, J. (1968). Kellen, W. R., Chapman, H. C., Clark, T. B., and Lindegren, J. E. (1965). Kellen, W. R., Chapman, H. C., Clark, T. B., and Lindegren, J. E. (1966). Kellen, W. R., Clark, T. B., and Lindegren, J. E. (1967). Kudo, R. (1962).
Microsporidian	*Indoplanorbis exustus* (*snail*)		Basch, P. F. (1971).

[a] Some of the examples cited under types C and D may be misclassified.

[b] Infections of artifically reared species where conditions of rearing separate generations and restrict intergeneration transfer to a vertical path alone and infections of univoltine species in which the agent overwinters in diapause eggs of the host.

classified in the type A category. These organisms, however, are classified elsewhere because the available quantitative information indicates that these agents are not selectively beneficial to their hosts (either directly or indirectly), and are not efficiently transmitted from males to their progeny. Thus, it is considered unlikely that any of the known microsporidia are maintained according to the type A pattern.

In type B, vertical transmission provides the sole means of transfer between generations of hosts, although alternative "horizontal" transfer occurs within the host generation. Such a transmission pattern is likely to occur only in species whose generations do not overlap in time. Non-overlapping of generations occurs in nature with univoltine species, which pass through a single generation per year and overwinter in the egg stage. In these examples the intra-generational horizontal transfer of infection is mostly likely to occur among larvae ingesting microbes shed by their congenitally infected peers. In some situations this transfer may occur between adults as in the transfer of Keystone virus in *Aedes atlanticus* (Fine and LeDuc, 1978). The segregation of generations implicit in type B transmission may also occur in some artificial situations (e.g., as a consequence of certain husbandry regimens for silkworms or for experimental insects in laboratory insectaries). As indicated by the diagram, vertical transmission is essential for the maintenance of type B infections in their host populations.

The type C transmission pattern represents a modification of the type B pattern. Horizontal transfer occurs both within and between overlapping generations during a portion of the year, but the infection is dependent entirely upon the vertical transmission pathway at some period when the entire population exists in the egg stage. Such a pattern may be common in the maintenance of pathogens affecting temperate zone insects which pass the winter as diapause ova. Again, the vertical transmission mechanism is qualitatively essential for infection maintenance, but there may be very extensive opportunity for horizontal transmission during part of the year, which can compensate for a drastic drop in the prevalence rate of infection because of an inefficient vertical transmission process (i.e., a low vertical transmission rate *vide infra*).

The type D transmission pattern represents what is probably the most common situation found in nature. Horizontal transmission occurs both within and between all generations of hosts and supplements the inter-generational vertical transfer. It is clear from the diagram that vertical transmission is not qualitatively essential for maintenance of the infection in these situations because of the availability of horizontal transfer between all generations. In this particular pattern there may be a quantitative requirement for vertical transmission in order to supplement an otherwise inadequate inter-generational horizontal transfer process. Whether or not the vertical transmission modality is indeed essential for the maintenance of a given type D infection in nature, or whether it serves only to increase prevalence rates above what the horizontal transmission alone could support are important problems. An argument appropriate for these considerations is discussed later.

The four patterns cited above provide a basic classification scheme for all vertically-transmitted infections of invertebrates. The classification could be expanded by the addition of subcategories with additional details (e.g., sex ratio distortions) but such an elaboration would compromise the simple logic illustrated here.

It should be emphasized that the examples cited in Table 2 are by no means an exhaustive catalogue of vertically-transmitted pathogens of invertebrates. They are meant only to illustrate the classification system and to provide a literature base for readers of this review. Readers should be cautioned that some examples may have been wrongly classified here, and others may belong to different transmission types under different environmental conditions. The author would appreciate corrections and suggestions for the improvement of this classification. Finally, given the current revisions of nomenclature, it is probable that some of the names cited in Table 2 have been superseded since their original publications.

III. GENERAL IMPLICATIONS OF THE VERTICAL TRANSMISSION OF PATHOGENS OF INVERTEBRATES

Recognizing that the vertical transfer of infectious agents is wisespread among invertebrates, several questions come to mind. What is its role in the ecology of these infections? It is an important phenomenon? If so, how and why? Is it worth studying, and if so, what are the questions one should be asking?

The invertebrate pathology literature contains many statements as to the role of vertical transmission. Examples of these statements, drawn from different sources cited within the bibliography of this paper, are given below:

1. "Data from infected progenies reared from field-collected eggs support the view that transovarian transmission of (microsporidian genus) is common in (insect host) and in many instances is sufficient to account for the levels of infection observed in the field."

2. With respect to a microsporidian "the very low rate of infection found in the wild mosquitoes is consistent with this view of the (per os) method of transmission. Were transovarian transmission usual, the rate of infection in wild mosquitoes should have been much higher."

3. "Because (microsporidian X) is transovarially transmitted, one would expect it to be more widespread and to infect a greater percentage of (insect host) colonies than either of the other two microsporidians (Y and Z)."

4. "As the generations pass...there is the possibility for the percentage of latently infected larvae or the level of infection which each larva carries to increase, due to slow multiplication and transovarial transmission of the virus."

5. "The efficiency with which the parasite is transmitted transovarially (at least under laboratory conditions) also serves to indicate the probable importance of this microsporidian as a natural mortality factor in field populations of (insect host)."

6. With respect to insect viruses: "If (transovarially) transmitting strains could be established, they might be valuable agents in biological control program."

Readers familiar with the literature on this subject will recognize such quotations as typical of views which have been widely expressed. Indeed, the statements represent the opinions of their authors, but each statement is of questionable validity in its assertion of direct correlations between vertical transmission and high or low field prevalence rates, field mortality rates, and biological control potential. It is this author's opinion that the several views expressed in the quotations above are logically unsound. Each reflects a simplistic view of the role of vertical transmission in the ecology of natural infections. More specifically, each statement fails to interpret the implications of vertical transmission with careful reference to the availability and role of alternative (horizontal) transmission pathways, the quantitative efficiency of the vertical transmission pathway itself, and the effect of infection upon survival and reproductive fitness of the host. The rest of this paper is devoted to discussing a rigorous, logical framework for analysis and an interpretation of this important issue as it relates to the epidemiological implications of vertical transmission mechanisms.

The simple classification and diagrams presented attempt to

make an initial step towards defining the importance of the vertical transmission of different pathogens of invertebrates. In three of the transmission patterns (types A, B, and C) the vertical transmission pathway is obviously essential for the perpetuation of the infectious agent in nature. On the other hand, there are major questions concerning the role of vertical transfer in the spatial distribution of infection, the precise mechanism for maintainenance by the extreme type A pattern, and the quantitative relationships between the horizontal and vertical transfer pathways in types B and C. I recognize that the majority of invertebrate infections probably fall within the type D category for which there is the possibility of horizontal transmission both within and between all host generations. In this type D there is thus no obvious qualitative role for vertical transfer and the impact of such transmission must be defined in terms of the relative quantitative extent of different transmission pathways.

IV. FRAMEWORK FOR QUANTITATIVE ARGUMENT

Interpretation of the epidemiological impact of a vertical transmission pattern raises important quantitative problems. A simple approach to such problems is presented below in the context of a general argument which is appropriate for analysis of the vertical transfer of any infection - mutual, commensal, or parasitic (= pathogen) - in any host population. Readers with general biological backgrounds will recognize close analogies between this approach and the so-called "Hardy-Weinberg equilibrium" argument that is basic to population genetics. As in the case of the Hardy-Weinberg argument, the simplicity of the assumptions is justified by the interest, generality, and robustness of the results.

We begin by specifying the parameters to be used with emphasis on those variables that can readily be measured in the laboratory. Then, we can specify the assumptions of the argument. Lastly, we discuss some of the logical outcomes of these assumptions with particular emphasis on the capabilities of vertical transmission mechanisms to maintain infectious agents in nature.

Five parameters are used:

p = Prevalence rate of infection among adult hosts of reproductive age.

d = Maternal vertical transmission rate = prevalence rate of infection among newly hatched (or live born) progeny of infected females mated with uninfected males. Note that this definition is independent of whether transmission is transovarial or transovum.

v = Paternal vertical transmission rate = prevalence rate of infection among newly hatched (or live born) progeny of uninfected females mated with infected males. Note that this definition is independent of the precise anatomical mechanism (i.e., it applies equally regardless of whether the transfer is direct via sperm to ovum or indirect via venereal transmission to the female). Such paternal-mediated vertical transmission has been recorded for several species of microsporidia (Thompson, 1958; Kellen and Lindegren, 1971), spirochaetes (Wagner-Jevseensko, 1958), rickettsia (Philip and Parker, 1933), and viruses (l'Heritier, 1970; Hamm and Young, 1974).

α = Relative fertility (number of progeny) of infected adults compared with their uninfected peers. This number represents the ratio of the offspring numbers produced by infected adults compared with uninfected adults.

β = Relative survival potential (to reproductive age) of congenitally infected young when compared with uninfected young. This number represents the ratio of the probabilities of survival to reproductive age of congenitally infected compared with uninfected young.

The product $\alpha\beta$ provides a measure of the total effect of a congenital infection on the reproductive success of its host and is closely analogous to the Darwinian fitness measure used by population geneticists in describing the effect of a gene upon its host. Each measures the relative reproductive success of infected (or carrier) individuals compared to uninfected or non-carrier individuals.

Each of these five parameters is readily measurable in any vertical transmission system in which hosts can be laboratory reared (see comment below concerning latency and diagnostic sensitivity). In order to use these parameters, one must specify the assumptions under which they are to be manipulated. There are essentially three assumptions to the general argument: (1) the infections are permanent in individual hosts (i.e., no "recoveries" occur; (2) the infection is similar (in prevalence rate p and in α and β) in both males and females of the host species. This assumption is obviously not applicable to all invertebrate infections; and (3) males and females of the host species mate at random without regard to the infection status of their partners. This assumption is necessary for the initial logic, but the results of the model apply even if ideal random mating does not occur in nature. Insofar as male-mediated vertical transmission is uncommon among invertebrate systems, this assumption is irrelevant, applying only when male-mediated transfer occurs. It is also irrelevant in parthenogenetic systems in which the model would apply fully by setting $v = 0$.

The results from these assumptions are cited below. The general algebraic solution is given in the appendix to this paper and have been discussed more completely in other publications (Fine, 1974, 1975).

(1) Assuming a type A transmission pattern (i.e., no horizontal transmission occurs), then specification for values of d, v, α, and β permits a prediction of the trend in prevalance rates of the infection over successive host generations. Figure 3 illustrates such predictions for five different sets of parameters.

(2) If an infection imparts a selective advantage to its host (i.e., it is a mutual, $\alpha\beta>1$), then it may be maintained in a host population by vertical transmission alone, even if $(d+v)<1$ and provided that the condition $\alpha\beta(d+v)>1$ is upheld.

(3) If an infection is a perfect commensal and has no effect at all upon its host (i.e., $\alpha\beta=1$), then it can only be maintained in the host population by vertical transmission alone and provided that $(d+v)>1$. If $\alpha=\beta=1$, the stable equilibrium prevalence rate of infection in the host population would be:

$$\frac{d+v-1}{dv}$$

(4) If an infection has a deterimental effect upon its host (i.e., a parasite with $\alpha<1$ or $\beta<1$, such that $\alpha\beta<1$), then a necessary condition for its maintenance by vertical transmission alone is $(d+v)>1$ and a sufficient condition is $\alpha\beta(d+v)>1$. This latter condition provides a useful criterion to assess whether any infection can be maintained within its host population by vertical transmission alone.

One can now apply these results to several issues commonly discussed in the invertebrate pathology literature:

(1) *Can pathogens of invertebrates be maintained by vertical transmission alone?* Several authors have argued that certain microsporidia or viruses are maintained or can be

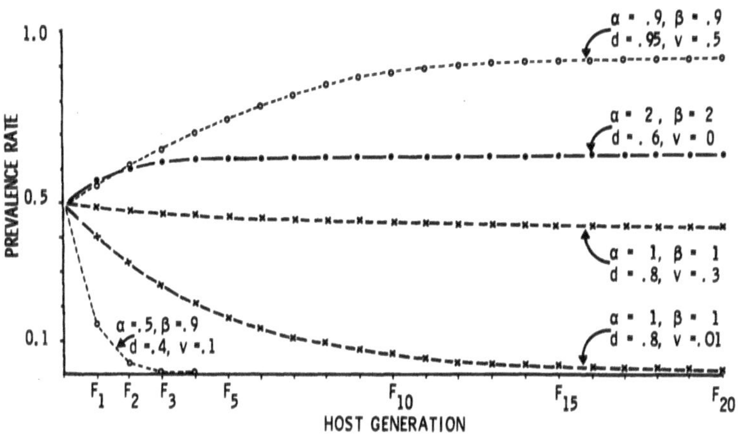

Figure 3. Predicted prevalence rates of infection among adults of successive generations of a host species, assuming type A transmission (i.e., only vertical transmission occurs). Assumes random mating, no sex differences, and no recoveries from infection. Parameters as defined in the text: d, v, = maternal and paternal vertical transmission rates; α, β = relative fertility and survival of infected compared to uninfected individuals. Prevalence rate in the initial ("P") generation set arbitrarily at 50%. Note that both pathogenic (αβ<1, lines o——o) and commensal (αβ=1, lines x——x) infections may be maintained by vertical transmission, but only if both rates are high enough that $\alpha\beta(d+v)>1$. Note also that a mutual (αβ>1, line •——•) may be maintained by vertical transmission from only a single parent, provided that $\alpha\beta(d+v)>1$.

maintained in their host populations by vertical transmission alone (see quotations in Section III). Consideration of the quantitative demands for such maintenance (e.g., result 4) reveals that maintenance by vertical transmission alone is rather unlikely. In effect it could occur only if both maternal and paternal vertical transmission rates were quite high, and if the pathogenicity were quite low, such that $\alpha\beta(\underline{d}+\underline{v})>1$. These conditions are probably not upheld by any of the common pathogens of invertebrates.

On the other hand, there are several well studied infections of Diptera which do deserve mention here. For example, the CO_2 sensitivity ("sigma") virus of *Drosophila melanogaster* is apparently maintained in fruit fly populations world-wide by vertical transfer alone (Seecof, 1968; L'Heritier, 1970). This maintenance may be achieved in spite of a slight selective disadvantage attributable to the infection (i.e. $\alpha\beta<1$), although this selective effect has not been carefully measured. Maintenance is achieved through the assistance of male as well as female-mediated vertical transmission, such that $\alpha\beta(\underline{d}+\underline{v})>1$.

The *Wolbachia* agents infecting mosquitoes of the *Culex pipiens* complex have evolved another mechanism for ensuring their maintenance by the type A transmission pattern. It has been shown that a cytoplasmic incompatibility results when an uninfected female mates with an infected male; however, an infected female mated with an uninfected male is compatible (Yen and Barr, 1973). The fact that uninfected females mated with infected males are sterile places the uninfected female, if infection is present in a population, at a disadvantage compared to an infected female, which is compatible regardless of her mate's pathology. Thus the *Wolbachia* has evolved an indirect density-dependent mechanism to ensure $\alpha>1$ and hence $\alpha\beta(d+v)>1$ (Fine, 1978a).

(2) *Can infection-dependent effects on sex ratio explain maintenance by vertical transmission alone?* It is well recognized that many vertically transmitted infections of invertebrates affect the sex ratio of their host populations. As a general rule, these infectious agents are selectively detrimental to males. The better known examples of such sex specific effects are several *Thelohania* microsporidia that infect mosquitoes (Kellen et al., 1965), sex ratio spirochaetes and other infectious agents associated with *Drosophila* (Poulson, 1963), and possibly *Rickettsia tsutsugamushi* that infect *Leptotrombidium* mites (Roberts et al., 1977). Thus one may inquire as to whether this factor, given that it is contrary to the first assumption of the basic argument, can account for the maintenance by vertical transfer alone. The answer is relatively straight forward: as long as no vertical transfer occurs via male parents (i.e. $v=0$, or else no infected adult males are present), then all of the conclusions are applicable, provied that the α and

β parameters are redefined so as to refer to female hosts only. Therefore, if paternally mediated vertical transmission of *Thelohania* does not occur, and it has not been reported, then the basic argument (result 4) applies and, given that $\alpha\beta d<1$, the prevalence rate must fall in successive generations in the absence of horizontal transmission.

Although the sex ratio altering effect of some microsporidia may not assure their transmission by the type A pattern, other sex ratio alterations may play an indirect role. An example is provided by the spirochaetal SR (Sex Ratio) factors of *Drosophila*. Carrier *Drosophila* females give rise to virtually no male progeny. As maintenance of SR spirochaetes is thought to be due the result of maternal transmission, with v being effectively nil, the argument indicates that the product $\alpha\beta d$ must exceed unity. Thus the SR agent must impart some selective advantage to its host (i.e., $\alpha\beta>1$). There is some direct evidence for this selective effect. It has been reported that infected females on the average give rise to more female offspring than do uninfected females. This result was first interpreted as evidence that the SR carrier agent somehow stimulated egg production in infected females and implied that $\alpha>1$ (Sakaguchi and Poulson, 1963). Later, this interpretation was modified when it was recognized that the numbers of ova produced by infected and uninfected females were approximately equal (implying $\alpha=1$). However, the female offspring of infected flies are more likely to survive than are the female offspring of uninfected flies (implying $\beta>1$) (Counce and Poulson, 1966). This improved survival rate may result from reduced competition for larval food in the absence of male larvae. Then, the sex ratio alteration effect could be indirectly responsible for changing the value of $\alpha\beta>1$, such that $\alpha\beta d>1$. This effect apparently permits the maintenance of these agents by vertical transmission alone (the type A pattern).

(3) *Is there a simple relationship between vertical transmission and prevelence rate of infection in a population?* Several authors mentioned in the previous section have implied that the occurrence of vertical transmission should be associated with high prevalence or mortality rates in a host population. There is little evidence to support this contention. The quantitative argument indicates that several variables, in addition to d, must be considered in assessing the impact of a vertical transmission mechanism on prevalence rates. The various SR agents of *Drosophila* spp. generally have high maternal vertical

transmission rates (d>0.9). In spite of such efficient vertical transfer, the prevalence rate of non-spirochaetal SR agents varies from 6 to 30% in wild populations of *D. bifasciata* in Italy (Poulson, 1963). Although actual surveys have not been taken, it appears that the spirochaetal SR agents of *D. willistoni* and *D. paulistorum* are present at very low prevalence rates (<1%) in the wild (Malogolowkin and Poulson, 1958). Clearly, considerable more information, in addition to the vertical transmission rate is needed for predicting levels of prevalence or mortality in a population.

(4) *Can horizontal transmission requirements be inferred from vertical transmission rates?* The quantitative argument outlined above permits an estimate of the contribution of a given vertical transmission mechanism (in terms of d, v, α,β) towards the maintenance of any given prevalence rate of infection in a population. Given a prevalence rate P in a parental generation, the fundamental equation (see appendix) provides an estimate of P', the predicted prevalence rate among adults in the progeny generation, provided that no horizontal transmission has occurred. If P' is less than P, then alternative transmission processes must be involved if the prevalence rate P is maintained in nature. Andreadis and Hall (1979b) have applied this analytical approach to data on the vertical transmission of a microsporidian, *Amblyospora* sp. in *Culex salinarius* based on parameter values of d=0.9, v=0, α=0.48, β=1.0. These investigators concluded that *Amblyspora* sp. would rapidly disappear from its host population if it were not for alternative means of transmission.

If a given prevalence rate P is known to be maintained in the field and the vertical transmission mechanism is not itself sufficient for such maintenance, then an estimate of the required horizontal transfer input may be obtained through the fundamental equation. Thus, if (P-P') measures the deficit in prevalence rate (which must be compensated by horizontal transmission during the progeny generation), the net incidence rate of horizontally acquired infections per generation is (P-P')/(1-P').

V. SUGGESTIONS FOR FURTHER RESEARCH

In general the literature indicates that further research needs to be directed towards a quantitative approach in studies on vertical transmission of invertebrate pathogens. More specifically, the approach should include the following suggestions:

(1) *More concentration upon population implications and less upon anatomical pathways.* Much discussion in the literature has been devoted to whether or not a vertically transmitted agent is transferred via transovarial or transovum transmission. In many cases this issue appears less important for understanding the agent's biology than does the issue of the extent to which transmission occurs. The important epidemiological factors include the successful transmission of the agent from the parent to its progeny, the parent involved in the transmission, and the frequency of transmission rather than the physical location of the agent on the egg.

There are situations in which the distinction between transovum[2] and transovarial transfer is useful. In some cases this distinction may be a useful characteristic in taxonomic description of infectious agents. In other cases, the distinction may provide a key to the control of infection in a laboratory population, insofar as transovum, but not transovarial, transmission may be blocked by appropriate surface sterilization of eggs (Alger and Undeen, 1950). If transovarial transfer is involved, control may require the selection and isolation-rearing of uninfected material from contamined stock.

Another important use for studies of anatomical sites of infection in host eggs, larvae, and imagos is for improving methods of diagnosis. Sensitive methods for recognition of infection are essential to the study of infections in populations. In the absence of these techniques investigators have been forced to measure fatalities-due-to-infection rather than the infection itself (David et al., 1968). However, fatalities represent an extreme outcome of the infection process and the number or rate generally constitute an inexact measure of the extent of actual infection. It is probable that the use

[2]The issue of "transovum" transmission is sometimes confused with "per os" transmission. The actual transfer of an infectious agent which is carried on the exterior of an egg may occur through ingestion (*per os*) by the hatching larva.

of crude measures of infection is in part responsible for the vague concept of "latency" found so commonly in the invertebrate pathology literature. Different authors have used this term to describe different phenomena which may only reflect results from the use of insensitive diagnostic technique. Studies aimed at improving diagnostic methods also should be encouraged in addition to studies on the site and ecology of infection.

(2) *Greater attention to quantitative data and appropriate study parameters.* There is little quantitative information available in the current literature on vertically transmitted infections of invertebrates. In the absence of such information, the implications of such transmission phenomena remain unclear. The five parameters (P, d, v, α, and β) defined earlier are recommended for use by all investigators so that the data generated are comparable, although other parameters may be found useful for special problems. For example, agents which differ in their effects on the fecundity of male and female hosts may require separate parameters for each factor. Workers who develop and use such parameters are urged to define them precisely. Also, when reporting experimental results investigators should include data on infected adults who fail to transmit vertically. Omission of these data lowers the denominator and leads to inflated estimates of vertical transmission rates.

The implications of the paternal vertical transmission rate parameter, v, also deserve discussion. As mentioned earlier it may be defined as the prevalence rate of infection among newly born or newly hatched progeny of uninfected females mated with infected males. It should be easily measurable by laboratory experimental systems as illustrated in the studies of Kellen and Lindegren (1971) on *Nosema plodiae*. The rate is independent of the invasive pathway (whether directly or indirectly to the ovum) as mentioned earlier. Given the importance of this rate for the maintenance capabilities of vertical transmission mechanisms, it is unfortunate that so little effort has been devoted to its measurement or to its implications.

(3) *Use of life table methods in presentation and analysis of data.* There is very little discussion of age or duration-of-infection in the invertebrate pathology literature. Age of the host may play an important part in determining the outcome and implications of an infection. Several authors have studied the effects of infections upon survivorship and/or fertility of their

hosts, but the different collection and reporting methods used has made the data difficult to compare (Anthony et al., 1978; Veber and Jasic, 1961; Undeen and Alger, 1975; Smirnoff and Chu, 1968). It is known that vertical transmission rates may vary with the age of the parent and with the duration of infection (Andreadis and Hall, 1979b; Singh et al., 1976; Sylvester, 1969). Collection of these data requires careful experimental protocols and may be simplified by the use of caged invertebrate populations. The standard life table is an appropriate format for presenting, analyzing, and interpreting such data. Conventional symbols such as lx (the probability of survival to age x) may be easily altered so as to describe the fate of both infected $(lx+)$ and uninfected $(lx-)$ individuals methods. Fine and Sylvester (1978) have described a method for deriving summary estimates of d, v, a, and β parameters for systems in which vertical transmission rates vary with age of host.

(4) *More emphasis on field studies and sampling.* Much of the published information on the vertical transfer of pathogens in invertebrates relates to laboratory observations. However, information on vertical transmission under field situations also is highly desirable. Special consideration must be given to sampling methodology since vertical transmission may predispose itself to spatial clustering of infection among the progeny of only a few infected females.

Field studies are important for elucidating the role of vertical transmission in pathogens that overwinter (e.g., distinguishing type B or type C transmission patterns) and in determining their spatial distribution (studies of the relationship between oviposition sites and infection foci in subsequent generations). Obviously, laboratory stocks represent the results of artifical selection from wild population material. This selection may have wild implications for studies of vertical transmission since several reports of laboratory selection for changes in vertical transmission rates are available (e.g., Geigy and Aeschlimann, 1964; Tesh, 1979). Thus every effort should be made to use fresh material and to relate laboratory results to field observations.

VI. CONCLUSIONS

There is a natural progression from descriptive to analytical emphasis in the evolution of a scientific discipline. Thus one may predict that knowledge of vertical transfer mechanisms of invertebrate pathogens will yield important benefits in the coming years. The literature now contains numerous anecdotes and descriptions of the phenomenon and initial efforts have been made to consider its significance in general terms. Further descriptive advances are imminent as recent findings concerning meiosis in microsporidia that congenitally infect male hosts are extended to the life cycle of these agents. Careful quantitative studies of vertical transmission, as illustrated by the recent work of Andreadis and Hall (1979b), should add greatly to our understanding of the epidemiological implications of vertical transmission in nature.

Also, in investigating vertical transmission of any pathogen it is important to realize that one is not probing the typical Mendelian style of genetics, but genes transmitted extrachromosomally. The subject of extrachromosomal genes and cytoplasmic heredity (and let us point out the difficulty in defining the dividing line between some hereditary intracellular infectious symbiotes such as *Wolbachia* and conventional cytoplasmic organelles such as mitochondria) has been one of the most rapidly expanding branches of biology in recent years (Fine, 1978b, 1979b). Two major discoveries have given rise to this growth: recognition of the presence of autonomous genetic continuity within such organelles as plastids and mitochondria, and recognition that vertical transmission plays an important role in the maintenance of many infectious agents at all levels of biological organization, including bacteria (e.g., lysogenic phages, episomes, plasmids protozoa (e.g., the many endosymbiotes of *Paramecium* spp.), invertebrates (e.g., mutuals and pathogens of insects), plants (e.g., seed-borne fungi and viruses) and vertebrates (e.g., the cancer viruses of rodents, cattle, and birds). The invertebrates, especially insects, have immense advantages over other life forms for use as laboratory models, (i.e., size, cheap and simple maintenance, rapid generation time, and high biotic potential). Studies with *Drosophila* have contributed considerably to our knowledge of vertical transmission, especially in our understanding of sigma, Sex Ratio (SR), *Mycoplasma*-like organisms (MLO), and related agents (L'Heritier, 1970; Poulson, 1963; Ehrman and Kernaghan, 1971). Other insect genera and species also have considerable potential in this field of endeavor.

Finally, in addition to emphasizing the basic scientific interest in the study of vertically transmitted infections, brief mention of the important practical implications of such research is warranted. A number of the pathogens which have been considered for biological control of insect pests are known to be vertically trasmitted. The implications of this transmission modality must be understood if one is to make accurate predictions of the impact of introducing a pathogen into a wild population. As discussed in this review, it is probable that several researchers have seriously misjudged these implications in the past because of a lack of quantitative data. It is recommended that Pasteur's (1870) classical work on the vertical transmission of *Nosema bombycis* can still be used as a model by researchers today.

VII. APPENDIX

Given the five parameters defined in the test and the assumptions (1) no horizontal transmission, (2) random mating, (3) no recoveries, and (4) equal effects on male and female hosts, the following results follow (for a complete discussion, see Fine, 1974): Defining P as the prevalence rate of infection among adults of a parent generation and P' as the prevalence rate of infection among adults of the progeny generation, then in general

$$P' = \frac{\beta\{P\alpha d\ (1-P+P\alpha) + P\alpha v\ (1-P+P\alpha\ -\ P\alpha d)\}}{\beta\{P\alpha d\ (1-P+P\alpha) + P\alpha v\ (1-P+P\alpha-P\alpha d)\} + (1-P+P\alpha-P\alpha d)\ (1-P+P\alpha-P\alpha v)}$$

If the hosts are parthenogenetic, or if no vertical transmission occurs from males (i.e., v=0), then

$$P' = \frac{P\alpha\beta d\ (1-P+P\alpha)}{P\alpha\beta d(1-P+P\alpha) + (1-P+P\alpha-P\alpha d)\ (1-P+P\alpha)}$$

If the infection has no effect on the host (i.e., $\alpha=\beta=1$, then

$$P' = P(d+v-Pdv)$$

Stable equilibrium prevalence rates of infection, which would ultimately be attained given any set of α, β, d, and v parameters, may be found either by iteration or else by substituting P for P' in the above equations and solving the resultant experessions for P. This procedure is discussed in Fine (1975).

VIII. REFERENCES

Alger, N. E. and Undeen, A. H. (1970). The control of a microsporidian, *Nosema* sp., in an anopheline colony by an egg rinsing technique. *J. Invertebr. Pathol.*, 15, 321-327.

Anderson, J. F. (1968). Microsporidia parasitizing mosquitoes collected in Connecticut. *J. Invertebr. Pathol.*, 11, 440-455.

Andreadis, T. G. and Hall, D. W. (1979a). Development, ultrastructure and mode of transmission of *Amblyospora* sp. (Microspora: thelohaniidae) in the mosquito *Culex salinarius*. *J. Protozool.*, 26, 444-452.

Andreadis, T. G. and Hall, D. W. (1979b). Significance of transovarial infections of *Amblyospora* sp. (microspora: thelohaniidae) in relation to parasite maintenance in the mosquito *Culex salinarius* Coquillett. *J. Invertebr. Pathol.*, 34, 152-157.

Anthony, D. W., Lotzkar, M. D., and Avery, S. W. (1978). Fecundity and longevity of *Anopheles albimanus* exposed at each larval instar to spores of *Nosema algerae*. *Mosq. News*, 38, 116-121.

Aruga, H. and Nagashima, E. (1962). Generation-to-generation transmission of the cytoplasmic polyhedrosis virus of *Bombyx mori* (Linnaeus). *J. Insect Pathol.*, 4, 313-320.

Basch, P. F. (1971). Transmission of microsporidian infection in the snail *Indoplanorbis exustus* in West Malaysia. *S. E. Asian J. Trop. Med. Publ. Hlth.*, 2, 380-383.

Bird, F. T. (1961). Transmission of some insect viruses with particular reference to ovarial transmission and its importance in the development of epizootics. *J. Insect Pathol.*, 3, 352-380.

Brooks, W. M. (1968). Transovarian transmission of *Nosema heliothidis* in the corn earworm *Heliothis zea*. *J. Invertebr. Pathol.*, 11, 510-512.

Brown, E. S. and Swain, G. (1965). Virus disease of the African armyworm, *Spodoptera exempta* (Wlk.). *Bull. Entomol. Res.*, 56, 95-117.

Buchner, P. (1965). "Endosymbiosis of Animals with Plant Microorganisms." *Interscience Publ.*, New York.

Bullock, H. R., Mangum, C. L., and Guerra, A. A. (1969). Treatment of eggs of the pink bollworm, *Pectinophora gossypiella*, with formaldehyde to prevent infection with a cytoplasmic polyhedrosis virus. *J. Invertebr. Pathol.*, 14, 271-273.

Burgdorfer, W. and Varma, M. G. R. (1967). Trans-stadial and transovarial development of disease agents in arthropods. *Ann. Rev. Entomol.*, 12, 347-376.

Canning, E. U. and Hulls, R. H. (1970). A microsporidian infection of *Anopheles gambiae* Giles, from Tanzania, interpretation of its mode of transmission and notes of *Nosema* infections in mosquitoes. *J. Protozool.*, 17, 531-539.

Chapman, H. C. 1974. Biological control of mosquito larvae. *Ann. Rev. Entomol.*, 19, 33-59.

Chapman, H. C., Woodard, D. B., Kellen, W. R., and Clark, T. C. (1966). Host-parasite relationship of *Thelohania* associated with mosquitoes in Louisiana (Nosematidae: Microsporidia). *J. Invertebr. Pathol.*, 8, 452-456.

Chapman, H. C. and Kellen, W. R. (1967). *Plistophora caecorum* sp. n., a microsporidian of *Culiseta inornata* (Diptera: culicidae) from Louisiana. *J. Invertebr. Pathol.*, 9, 500-502.

Chapman, H. C., Clark, T. B., Petersen, J. J., and Woodard, D. B. (1969). A two-year study of pathogens and parasites of Culicidae, Chaoboridae, and Ceratopogonidae in Louisiana. *Proc. N. J. Mosq. Exterm. Assoc.*, 56, 203-212.

Chapman, H. C., Woodard, D. B., and Petersen, J. J. (1967). Pathogens and parasites of Louisiana Culicidae and Chaoboridae. *Proc. N. J. Mosq. Exterm. Assoc.*, 54, 54-60.

Clark, T. B. and Fukuda, T. (1971). Field and laboratory observations of two viral diseases in *Aedes sollicitans* (Walker) in Southwestern Louisiana. *Mosq. News*, 31, 193-199.

Counce, S. J. and Poulson, D. F. (1966). The expression of maternally-transmitted sex ratio condition (SR) in two strains of *Drosophila melanogaster*. *Genetica*, 37, 364-390.

David, W. A. L., Gardiner, B. O. C., and Clothier, S. E. (1968). Laboratory breeding of *Pieris brassicae* transmitting a granulosis virus. *J. Invertebr. Pathol.*, 12, 238-244.

David, W. A. L. and Taylor, C. E. (1967). Transmission of granulosis virus in the eggs of a virus-free stock of *Pieris brassicae*. *J. Invertebr. Pathol.*, 27, 71-75.

Doane, C. C. (1969). Trans-ovum transmission of a nuclear-polyhedrosis virus in the gypsy moth and the inducement of virus susceptibility. *J. Invertebr. Pathol.*, 14, 199-210.

Doane, C. C. (1970a). Primary pathogens and their role in the development of an epizootic in the gypsy moth. *J. Invertebr. Pathol.*, 15, 21-33.

Doane, C. C. (1970b). Tranovum transmission of nuclear polyhedrosis virus in relation to disease in gypsy moth populations. *Proc. 4th Inter. Cong. Insect Pathol.*, 285-291.

Doane, C. C. (1975). Infectious sources of nuclear polyhedrosis virus persisting in natural habitats of the gypsy moth. *Environ. Entomol.*, 4, 392-394.

Ehrman, L. and Kernaghan, R. P. (1971). Microorganismal basis of infectious hybrid male sterility in *Drosophila paulistorum*. *J. Hered.*, 62, 66-71.

Etzel, L. K. and Falcon, L. A. (1976). Studies of transovum and transstadial transmission of a granulosis virus of the codling moth. *J. Invertebr. Pathol.*, 27, 13-26.

Fine, P. E. M. (1974). The epidemiological implications of vertical transmission. Ph.D. Dissertation. University of London, London, England.

Fine, P. E. M. (1975). Vectors and vertical transmission: an epidemiologic perspective. *Ann. N. Y. Acad. Sci.*, 266, 173-194.

Fine, P. E. M. (1978a). On the dynamics of symbiote-dependent cytoplasmic incompatibility in culicine mosquitoes. *J. Invertebr. Pathol.*, 30, 10-18.

Fine, P. E. M. (1978b). Mitochondrial inheritance and disease. *Lancet*, 2, 659-662.

Fine, P. M. M. (1979a, 1981). Epidemiological principles of vector mediated transmission. *In* "Vectors of Disease Agents: Interactions with Plants, Animals and Man." (J. J. McKelvey, B. F. Eldridge, and K. Maramorsch, eds.) pp. 77-91. Praeger, N.Y.

Fine, P. E. M. (1979b). Lamarckian ironies in contemporary biology. *Lancet*, 1, 1181-1182.

Fine, P. E. M. and LeDuc, J. W. (1978). Towards a quantitative understanding of the epidemiology of Keystone virus in the eastern United States. *Am. J. Trop. Med. Hyg.*, 27, 322-338.

Fine, P. E. M. and Sylvester, E. S. (1978). Calculation of vertical transmission rates of infection, illustrated with data on an aphid-borne virus. *Am. Naturalist*, 112, 781-786.

Fox, R. M. and Weiser, J. (1959). A microsporidian parasite *Anopheles gambiae* in Liberia. *J. Parasit.*, 45, 21-30.

Gaugler, R. R. and Brooks, W. M. (1975). Sublethal effects of infection by *Nosema heliothidis* in the corn earworm, *Heliothis zea*. *J. Invertebr. Pathol.*, 26, 57-63.

Geigy, R. and Aeschlimann, A. (1964). Langfristige Beobachtungen über transovarielle Übertragung von *Borrelia duttoni* durch *Ornithodorus moubata*. *Acta Trop.*, 21, 87-91.

Gorske, S. F. and Maddox, J. V. (1978). A microsporidium, *Nosema pilicornis sp. n.*, of the purslane sawfly, *Schizocerella pilicornis*. *J. Invertebr. Pathol.*, 32, 235-243.

Grace, T. D. C. (1958). Induction of polyhedral bodies in ovarian tissues of the tussock moth *in vitro*. *Science*, 128, 249-250.

Gross, L. (1944). Is cancer a communicable disease? *Cancer Res.*, 4, 293-303.

Hall, D. W. and Anthony, D. W. (1971). Pathology of a mosquito iridescent virus (MIV) infecting *Aedes taeniorhynchus*. *J. Invertebr. Pathol.*, 18, 61-66.

Hamm, J. J. and Young, J. R. (1974). Mode of transmission of nuclear-polyhedrosis virus to progeny of adult *Heliothis zea*. *Invertebr. Pathol.*, 24, 70-81.

Harpax, I. and Ben Shaked, Y. (1964). Generation-to-generation transmission of the nuclear polyhedrosis virus of *Prodentia litura* (Fabricius). *J. Insect Pathol.*, 6, 127-130.

Hazard, E. I. and Weiser, J. (1968). Spores of *Thelohania* in adult female *Anopheles:* development and transovarial transmission, and redescriptions of *T. legeri* Hesse and *T. obesa* Kudo. *J. Protozool.*, 15, 817-823.

l'Heritier, P. (1970). *Drosophila* viruses and their role as evolutionary factors. *Evolutionary Biol.*, 4, 185-209.

Hukuhara, T. (1962). Generation to generation transmission of the cytoplasmic polyhedrosis virus of the silkworm, *Bombyx mori* (Linnaeus). *J. Insect Pathol.*, 4, 132-135.

Kellen, W. R. (1962). Microsporidia and larval control. *Mosq. News*, 22, 87-95.

Kellen, W. R. and Wills, W. (1962). The transovarian transmission of *Thelohania californica* Kellen and Lipa in *Culex tarsalis* Coquillett. *J. Insect Pathol.*, 4, 321-326.

Kellen, W. R., Chapman, H. C., Clark, T. B., and Lindegren, J. E. (1965). Host-parasite relationships of some *Thelohania* from mosquitoes (Nosematidae: Microsporidia). *J. Invertebr. Pathol.*, 7, 161-166.

Kellen, W. R., Chapman, H. C., Clark, T. B., and Lindegren, J. E. (1966). Transovarial transmission of some *Thelohania* (Nosematidae: Microsporidia) in mosquitoes of California and Louisiana. *J. Invertebr. Pathol.*, 8, 355-359.

Kellen, W. R., Clark, T. B., and Lindegren, J. E. (1967). Two previously undescribed *Nosema* from mosquitoes of California (Nosematidae: Microsporidia). *J. Invertebr. Pathol.*, 9, 19-25.

Kellen, W. R. and Lindegren, J. E. (1971). Modes of transmission of *Nosema plodiae* Kellen and Lindegren, a pathogen of *Plodia interpunctella* (Hubner). *J. Stored Prod. Res.*, 7, 31-34.

Kellen, W. R. and Lindegren, J. E. (1973). Transovarian transmission of *Nosema plodiae* in the Indian-meal moth, *Plodia interpunctella*. *J. Invertebr. Pathol.*, 21, 248-254.

Krinsky, W. L. (1977). *Nosema parkeri*, sp. n., a microsporidian from the argasid tick *Ornithodorus parkeri* Cooley. *J. Protozool.*, 24, 52-56.

Kudo, R. (1962). Microsporidia in Southern Illinois mosquitoes. *J. Insect Pathol.*, 4, 353-356.

Linley, J. R. and Nielsen, H. T. (1968a). Transmission of a mosquito iridescent virus in *Aedes taeniorhynchus*. I. Laboratory experiments. *J. Invertebr. Pathol.*, 12, 7-16.

Linley, J. R. and Nielsen, H. T. (1968b). Transmission of a mosquito iridescent virus in *Aedes taeniorhynchus*. II. Experiments related to transmission in nature. *J. Invertebr. Pathol.*, 12, 17-24.

Malogolowkin, G. and Poulson, D. F. (1958). Sex ratio in *D. willistoni*. *Proc. 10th Int. Cong. Genet.*, 2, 176.

Martignoni, M. E. and Milstead, J. E. (1962). Trans-ovum transmission of the nuclear polyhedrosis virus of *Colias eurytheme* Boisduval through contamination of the female genitalia. *J. Insect Pathol.*, 4, 113-121.

Nielson, M. M. and Elgee, D. E. (1968). The method and role of vertical transmission of a nucleopolyhedrosis virus in the European sawfly, *Diprion hercyniae*. *J. Invertebr. Pathol.*, 12, 132-139.

Nordin, G. L. and Maddox, J. V. (1974). Microspridian of the fall webworm, *Hyphantria cunea*. I. Identification, distribution, and comparison of *Nosema* sp. with similar *Nosema* spp. from other Lepidoptera. *J. Invertebr. Pathol.*, 24, 1-13.

Pasteur, L. (1870). *Etudes sur la Maladie des Vers à Soie*. Oeuvres de Pasteur, reunies par Pasteur Vallery-Radot. Masson et Cie., Paris (1926).

Philip, C. B. and Parker, R. R. (1933). Rocky Mountain spotted fever: investigation of sexual transmission in the wood tick *Dermacentor andersoni*. *Publ. Hlth. Rep.*, 48, 266-272.

Poulson, D. F. (1963). Cytoplasmic inheritance and hereditary infections in *Drosophila*. *In* "Methodology in Basic Genetics" (W. J. Burdette, ed.), Holden-Day, Inc. San Francisco.

Reinganum, C., O'Loughlin, G. T., and Hogan, T. W. (1970). A nonoccluded virus of the field crickets *Teleogryllus oceanicus* and *T. commodus* (Orthoptera: gryllidae). *J. Invertebr. Pathol.*, 16, 214-220.

Richardson, J., Sylvester, E. S., Reeves, W. C., and Hardy, J. L. (1974). Evidence of two inapparent nonoccluded viral infections of *Culex tarsalis*. *J. Invertebr. Pathol.*, 23, 213-224.

Roberts, L. W., Rapmund, G., and Cadigan, F. C. (1977). Sex ratios in *Rickettsia-tsutsugamushi*-infected and noninfected colonies of *Leptotrombidium* (Acari: Trombiculidae). *J. Med. Entomol.*, 14, 89-92.

Sakaguchi, B. and Poulson, D. F. (1963). Interspecific transfer of the "sex ratio" condition from *Drosophila willistoni* to *D. melanogaster*. *Genetics*, 48, 841-861.

Schwabe, C. W., Riemann, H. P., and Franti, C. E. (1977). *In* "Epidemiology in Veterinary Practice". Lea and Febiger, Philadelphia.

Seecof, R: L. (1968). The sigma virus infection *Drosophila melanogaster*. *Curr. Topics Microbiol. Immunol.*, 42, 59-93.

Sikorowski, P. P. and Madison, C. H. (1968). Host parasite relationships of *Thelohania corethrae* (Nosematidae, Microspiridia) from *Chaoborus astictopus* (Diptera: Chaoboridae). *J. Invertebr. Pathol.*, 11, 390-397.

Sikorowski, P. P., Andrewes, G. L., and Broome, J. R. (1973). Trans-ovum transmission of a cytoplasmic polyhedrosis virus of *Heliothis virescens* (Lepidoptera: Noctuidae). *J. Invertebr. Pathol.*, 21, 41-45.

Singh, K. R. P., Curtis, C. F., and Krishnamurthy, B. S. (1976). Partial loss of cytoplasmic incompatibility with age in males of *Culex fatigans*. *Ann. Trop. Med. Parasitol.*, 70, 463-466.

Smirnoff, W. A. (1961). A virus disease of *Neodiprion swainei* Middleton. *J. Insect Pathol.*, 3, 29-46.

Smirnoff, W. A. (1962). Trans-ovum transmission of virus of *Neodiprion swainei* Middleton (Hymenoptera, Tenthredinidae). *J. Insect Pathol.*, 4, 192-200.

Smirnoff, W. A. and Chu, W. H. (1968). Microsporidian infection and the reproductive capacity of the larch sawfly, *Pristophora erichsonii*. *J. Invertebr. Pathol.*, 12, 388-390.

Steinhaus, E. A. and Martignoni, M. E. (1970). *An Abridged Glossary of Terms Used in Invertebrate Pathology*. 2nd edition. Pacific Northwest Forest and Range Experiment Station, U.S. Dept. Agric.

Swaine, G. (1966). Generation-to-generation passage of the nuclear polyhedral virus of *Spodoptera exempta* (Wlk.). *Nature*, 210, 1053-1054.

Sylvester, E. S. (1969). Evidence of transovarial passage of the sowthistle yellow vein virus in the aphid *Hyperomyzus lactucae*. *Virology*, 38, 440-446.

Szollosi, D. (1971). Development of *Pleistophora* sp. (Microsporidian) in eggs of the polychaete *Armanda brevis*. *J. Invertebr. Pathol.*, 18, 1-15.

Tesh, R. (1979). Vertical transmission of arthropod-borne viruses of vertebrates. *In* "Vectors of Disease Agents: Interactions with Plants, Animals and Man." (J. J. McKelvey, B. F. Eldridge, and K. Maramorsch, eds.). pp. 122-137. Prager, New York.

Thompson, H. M. (1958). Some aspects of the epidemiology of a microsporidian parasite of the spruce budworm, *Choristoneura fumiferana* (Clem.). *Can. J. Zool.*, 36, 309-316.

Undeen, A. H. and Alger, N. E. (1975). The effect of the microsporidian, *Nosema algerae*, on *Anopheles stephensi*. *J. Invertebr. Pathol.*, 25, 19-24.

Urquhart, F. A. (1966). Virus-caused epizootic as a factor in population fluctuations of the monarch butterfly. *J. Invertebr. Pathol.*, 8, 492-495.

Vail, P. V. and Hall, I. M. (1969). The influence of infections of nuclear polyhedrosis virus on adult cabbage loopers and their progeny. *J. Invertebr. Pathol.*, 13, 358-370.

Vail, P. V. and Gough, D. (1970). Effects of cytoplasmic polyhedrosis virus on adult cabbage loopers and their progeny. *J. Invertebr. Pathol.*, 15, 397-400.

Van der Plank, J. E. (1963). "Plant Diseases: Epidemics and Control". Academic Press, New York.

Veber, J. and Jasic, J. (1961). Microsporidia as factors reducing the fecundity of insects. *J. Insect Pathol.*, 3, 103-111.

Wagner-Jevseensko, O. (1958). Fortpflanzung bei *Ornithodorus moubata* und genitale Ubertragung von *Borrelia duttoni*. *Acta Trop.*, 15, 118-168.

Woodard, D. B. and Chapman, H. C. (1968). Laboratory studies with the mosquito iridescent viurs (MIV). *J. Invertebr. Pathol.*, 11, 296-301.

Yen, J. H. and Barr, A. R. (1973). The etiological agent of cytoplasmic incompatibility in *Culex pipiens*. *J. Invertebr. Pathol.*, 22, 242-250.

Zimmack, H. L., Arbuthnot, K. D., and Brindley, T. A. (1954). Distribution of the European corn borer parasite *Perezia pyraustae* and its effect on the host. *J. Econ. Entomol.*, 47, 641-645.

Zimmack, H. L. and Brindley, T. A. (1957). The effect of the protozoan parasite *Perezia pyraustae* Paillot on the European corn borer. *J. Econ. Entomol.*, 50, 637-640.

EPIZOOTIOLOGY OF HAPLOSPORIDAN DISEASES AFFECTING OYSTERS

J.D. Andrews

Virginia Institute of Marine Science
and
College of William and Mary
Gloucester Point, Virginia 23062

I. INTRODUCTION

The first haplosporidan pathogen of oysters was found in 1957 in Delaware Bay (Haskin et al., 1966). Since then pathogens of this sporozoan group have been found in oysters on most continents. Five species belonging to two genera, each distinctive and confined to one coast, have been studied rather intensively. Four of these species are known to cause persistent diseases in populations of oysters. Summary descriptions of oyster mortalities have been presented by Sindermann (1976). Other haplosporidans of uncertain geographic origin and importance have been described from mussels and oysters, often from only a few specimens (Sprague, 1970; Kern, 1976).

The state of knowledge of haplosporidan diseases is deficient in several crucial aspects: (1) it is not known whether the pathogens are indigenous or exotic; (2) the sources of infections and the reservoirs of infective particles are unknown; (3) the infective stages have not been demonstrated or recognized; (4) the life cycles are incompletely known because infections have not been induced experimentally in the laboratory; (5) sporulation is rare or obscured in some species and is often erratic, which renders uncertain the normal life cycle of the pathogen in oysters and its possible alternate host; (6) alternate hosts have not been demonstrated for haplosporidans; (7) the natural hosts may be other species of bivalves or they may be other invertebrates, and (8) control of a disease organism usually requires a complete knowledge of its life cycle, particularly if an alternate host is involved.

The sudden explosive outbreaks of oyster mortalities in North America and Europe and the persistence of diseases suggest that new pathogens may have been introduced. All are essentially warm season diseases, but infection, sporulation, and subsequent mortality tend to be irregular. The irregular life cycle of the pathogen and the high mortality of its host indicate that the host is not well-adapted to the pathogen. The various diseases appear to have become endemic and the oyster populations are slowly adapting through intensive selection. Control of diseases in tidal marine waters is difficult to achieve by classical methods, i.e., interruption of the pathogen life cycle and reduction in the availability of hosts. Present research on control of oyster diseases is limited to breeding of resistant strains of oysters.

II. GENERAL CONSIDERATIONS AND CHARACTERISTICS

In this report I will attempt to characterize haplosporidan

diseases by the classical epizootiological categories (Frost, 1941). My approach will be directed primarily at studies of oyster populations with less emphasis placed on the traits of diseases expressed in individuals. Much of the literature on oyster diseases is concerned with fine structure and classification of pathogens, and the pathology of individual oysters. Quantitative descriptions of population aspects of epizootiology are limited primarily to levels of infection and mortality in oysters and to annual variations in activity of a disease.

The literature was found to be inadequate for characterizing European and Australian haplosporidans in most epizootiological categories. However, these diseases are briefly described and where appropriate information is available, such is summarized in Table 1. It is noteworthy that the three *Minchinia* diseases of oysters cause systemic infections, whereas the *Marteilia* diseases primarily effect the digestive tract. Other species of haplosporidans parasitize the digestive tract of certain snails (Barrow, 1965), boring molluscs (Hillman, 1979), mud crabs (Mackin and Loesch, 1955), and mussels (Taylor, 1966). Discoloration of the digestive tract by mature spores is an important diagnostic sign for detecting infections. All the oyster pathogens, except *Minchinia nelsoni*, sporulate regularly but the site varies between the epithelium of digestive tubules and connective tissues of all organs.

Table 1 also lists two diseases relatively new to Western Europe (primarily France and Spain) which are caused by haplosporidans. *Marteilia refringens* attacks the flat oyster, *Ostrea edulis* in Brittany (Marteil, 1976). The disease caused by it occurs in the digestive tract, appearing in August and causes death of the flat oyster the next year because of starvation (Balouet, 1979). The second haplosporidan, *Minchinia amoricana* (see Van Banning, 1979) is rarely found in Western Europe infecting *Ostrea edulis*. Its close relationship to *M. costalis* indicates that if introduced to North America it could cause significant mortality in species of *Crassostrea* such as the American oyster. The spores of *M. amoricana* are similar in size to spores of *M. costalis*. Both are systemic pathogens that sporulate in the connective tissues of oysters. However, observation by electron microscope of spore wrappings and their extensions indicate the two organisms to be different species (Van Banning, 1977; Perkins, 1968, 1979; Rosenfield et al., 1969; Perkins, pers. commun.). Both diseases appeared in France immediately after *C. gigas* was introduced from Japan in 1966,

TABLE 1. Disease characteristics of haplosporidan pathogens of oysters.

Name	Host	Period of Infectivity	Organs Infected	Period of Mortality	Time	Sporulation Site	Regularity
Minchinia nelsoni	*Crassostrea virginica*	May-Oct.	all (systemic)	all year	June (irr.)	epithelium of digestive tubules	rare
M. costalis	*C. virginica* (Mid-Atlantic Coast)	June-July	all (systemic)	May-June	June	connective tissues	regular
M. amoricana	*Ostrea edulis* (Western Europe)	?	all (systemic)	?	?	connective tissues	regular
M. pickfordae	freshwater snails (3 genera, 5 species) (Great Lakes, USA)	summer	digestive tract	summer	summer	epithelum of digestive tubules	regular
Marteilia refringens	*O. edulis* (Western Europe)	summer	digestive tract	late summer	summer	epithelium of digestive tubules	regular
M. sydneyi	*C. commercialis*	all year	digestive tract	all seasons, variable	all year	epithelium of digestive tubules	regular

lending credence to Marteil's (1968) hypothesis that the pathogens originated in Asia.

Marteilia sydneyi, a pathogen of *Crassostrea commercialis* in Australia, causes mortalities of oysters at irregular intervals. *M. sydneyi* causes a disease of the digestive tract similar to its close relative, *Marteilia refringens*. Epizootics occur throughout the year, presumably because of the relatively warm temperatures (>14°C) (Wolf, 1979).

Minchinia spp. have been reported (though rarely in the Pacific oyster, *C. gigas)* in Korean and Japanese oyster beds, but they have not been linked to population declines (Kern, 1976). The data collected in Asia is inadequate to allow comparisons with species of pathogens on other continents.

Table 2 gives a comparison of the epizootiology of three oyster diseases and their characteristics. The oyster pathogen *Perkinsus marinus* is included to emphasize the differences in epizootiology and pathogenicity of a disease that is more thoroughly understood. The other two disease organisms, *Minchinia nelsoni* and *M. costalis*, are now endemic to the Chesapeake Bay. *Minchinia nelsoni* (acronym, MSX) occurs along the mid-Atlantic coast of North America (Haskin et al., 1966). It is destructive from Chesapeake Bay to Long Island Sound, although it has been found from North Carolina to Massachusetts. The pathogen is present in medium- to high-salinity waters (>15 o/oo) throughout the year and primarily kills oysters in the summer and early fall. Up to 50 to 60% of previously unexposed susceptible oysters are killed annually.

The disease caused by *M. costalis* (acronym, SSO) is restricted to high salinity waters (>25 o/oo) along the coast from Virginia to New York (Wood and Andrews, 1962). It annually kills 30 to 50% of native oysters on the seaward-side of the Virginia-Maryland-Delaware coast.

The annual cycles of activity of these two diseases in the Chesapeake Bay also are illustrated by Figs. 1 and 2. Duplicate lots of susceptible James River oysters were imported and given full exposure to the disease organisms. Mortality caused by MSX in 1978 was slightly below average over a 20-year period. Mortality caused by SSO also was below average and some interference by MSX in June was noted. In most years on the seaward-side of the Eastern Shore little mortality was caused by MSX as occurred in the late summer of 1978 (Fig. 2).

TABLE 2. Epizootiology of oyster diseases in the Chesapeake Bay.

DISEASE AGENT	*Perkinsus marinus*	*Minchinia nelsoni*	*Minchinia costalis*
Winter survival	Low	High	High
Expulsion by oysters	November–May	Any time > 10o/oo or > 30o/oo	May–June known, other?
Age susceptibility	1–2 Years–Low 3–5 Years–High	1 Year–Low 2–4 Years–High	1 Year–Low 2–3 Years–High
Selection for resistance	Very difficult	Easy-Fast[2]	Probable to Moderate
Clinical signs (macroscopic)	Non specific	Non specific	Non specific
Alternate host	None	None known	None known
Period of infectivity	July–October	May–October	June–July
Level of infectivity	Low	High	High
Incubation period	1 month	1–10 months	10 months
Diagnosis method	Thioglycollate culture	Histological section	Histological section
Prevalence levels[b]	Up to 100% (variable)	30–60% (uniform)	20–50% (uniform)

DISEASE AGENT	*Perkinsus marinus*	*Minchinia nelsoni*	*Minchinia costalis*
Morbidity (annual)[b]	Very high (up to 100%)	High[c] (40-80%)	Moderate (20-50%)
Mortality (annual)	High[d] (up to 50%)	High (40-70%)	High (20-50%)
Classification	Apicomplexa	Haplosporida	Haplosporida
Geographical distribution	Delaware Bay South on the Atlantic Coast	Massachusetts to North Carolina	Maine to Virginia
Rapidity of dispersal	Slow	Fast	Fast
Salinity tolerance	12-15 to 35°/oo	15-30°/oo	>25°/oo

[a]High resistance achieved in native and laboratory stocks.

[b]In fully endemic areas. Rates in marginal zones may be lower.

[c]Involves two infection periods with continuous mortalities between late summer and spring peaks of prevalences.

[d]Higher in warmer climates with longer seasons above 20°C.

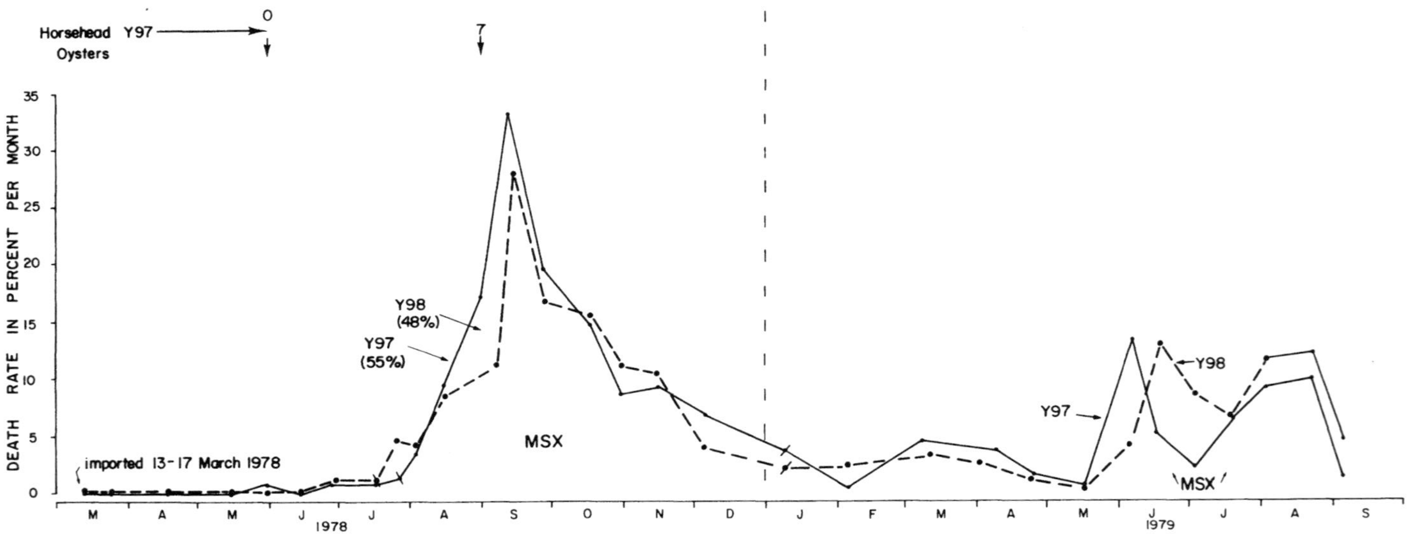

Figure 1. Patterns of *Minchinia nelsoni* (MSX) infection and mortality in transplanted, susceptible James River seed oysters. Mortality begins typically about 1 August each year in winter and spring transplanted oysters. Total losses for the warm season (between hash marks) are given with tray number. Prevalences of infection are given in number of cases per 25 live oysters above arrows designating sample dates.

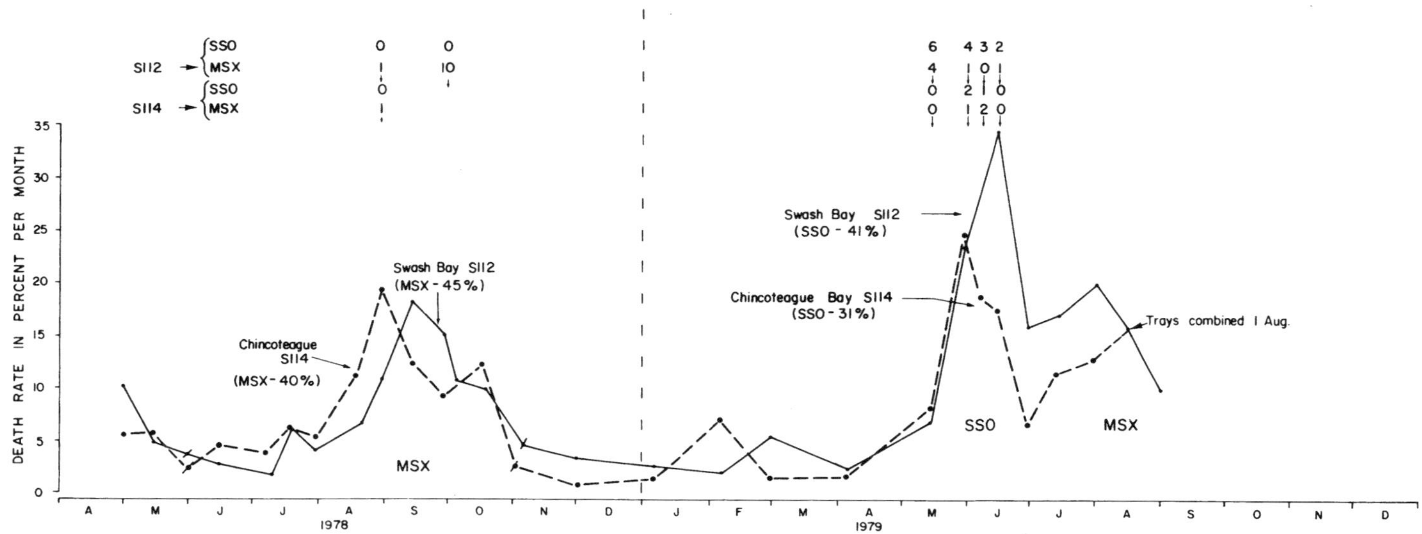

Figure 2. Patterns of infection and mortality for *Minchinia costalis* (SSO) on seaside of Eastern Shore Virginia. Replicate trays of James River seed oysters were monitored in two bays behind barrier islands. Prevalences of both diseases are given by dates of sampling. SSO does not kill oysters until a year after infections occur in June-July. Oysters exposed to MSX in June begin dying in August.

III. Epizootiology

The epizootiology of the diseases caused by *P. marinus*, *M. nelsoni*, and *M. costalis* as presented below is based on 20 years of continuous monitoring of oyster populations in trays and on natural beds.

Minchinia nelsoni and M. costalis. Geographical Distribution of *Minchinia nelsoni* and *M. costalis*. *M. nelsoni* appeared in Delaware Bay in the spring of 1957 (Haskin and Ford, 1979) and two years later it caused a summer population decline in Mobjack Bay and the lower Chesapeake Bay (Andrews and Wood, 1967). In each estuary the disease spread rapidly throughout the area. Salinity is a major factor limiting the distribution of the organism (Andrews, 1979). The pathogen prefers salinities between 15 and 25°/oo during the summer infective period. Remission of infection occurs at higher and lower salinities, particularly in the spring. MSX spreads throughout most oyster-growing areas of Delaware Bay, but is limited mostly to the Virginia side of the Chesapeake Bay and to the lower sectors of the major river tributaries in Virginia. Consequently, major populations of oysters in Chesapeake Bay were not exposed to infections of the disease because low salinity areas provided sanctuaries from the disease. These "safer" areas include the major seed areas in Virginia's James River. Thus, brood stocks in these sanctuaries produce susceptible seed oysters that should not be planted in high-salinity waters without excessive population losses (80-90% within two or three years).

Salinities above 25°/oo in Long Island Sound and New England as well as in the Carolinas and Georgia, may account, in part, for the failure of MSX to cause significant mortality in regions north and south of the mid Atlantic coast. Although *M. costalis* is believed to have caused serious mortality in Long Island Sound in 1953, it has not caused regular losses along the coast north of Cape Henlopen on the Delaware Bay or south of Cape Charles on the Chesapeake Bay.

Neither high nor low temperatures have been shown to suppress or eliminate *M. costalis* or *M. nelsoni*. The pathogens have persisted throughout the winter and have infected oysters during the summer regularly for 20 years.

Seasonal Distribution and Incubation Periods. Plasmodial infections of *M. nelsoni* are found in oysters throughout the year in areas where the disease is endemic. The pathogen is capable of long incubation periods (after late summer and fall infections) in areas where salinities are marginal for development of the disease. Infections acquired in August through October often remain subclinical until the following May. Thus, oysters for transplantation cannot be certified free of disease through most of the normal transplanting season (October through May) in Virginia. Any sampling program designed to monitor infections during the year must consider the incubation period of the pathogen.

M. costalis has a regular 8- to 9-month incubation period and cannot be diagnosed by stained sections of oyster tissues. Scarce uninucleate cells in the epithelial layer of digestive tubules are believed to represent latent infections.

Reservoirs of Infective Particles. Failure to induce laboratory infections of haplosporidan diseases in North America and Europe has led to widespread belief that hosts other than oysters are the sources of infective particles. Most marine parasites having alternate hosts require rather close proximity to the host for insuring transmission of the infection. However, oysters filter such large quantities of water that this requirement may not be essential. Oysters in trays that have been isolated from natural oyster populations acquire infections at the same time and rate as oysters in beds surrounded by natural communities.

Presumably, infective particles are filtered from the water by oysters. Only microscopic food organisms are permitted to pass into the branchial chamber. If infective particles are carried by plantonic organisms, one would expect most infections to occur in the digestive tubules after ingestion. In the case of *M. costalis* and the two species of *Marteilia* mentioned early, infections are found in the epithelium of tubules. However, most initial infections of MSX are found in the gill epithelium. Early infections are localized in the gills, suggesting that infective particles are relatively scarce but highly infective.

The efficiency of MSX in establishing endemic infections in entire populations of oysters is impressive. If an alternate host provides the infective particles one would suspect a mobile species, such as the blue crab, that is capable of discharging spores near the molluscan host. Unless there is some attraction to oysters, there likely is no reason for a parasite to use another host. Even assuming infections are produced by their near proximity to oysters, such an alternate host requires exceptional distribution to infect entire populations of oysters. Perhaps plasmodia are carried passively in the guts of scavengers, e.g., mud crabs, small fishes, after feeding on decaying oyster meat, but these animals are mostly local inhabitants of oyster beds and may lack the capability of wide dispersal in an estuary.

Annual fluctuations in the geographic range of MSX infections occur with changes in salinity regimes in the estuaries. Thus, infective particles may be carried tens of miles either by the hosts or by tidal currents. The dilution factor must become excessively large, but oysters filter up to 400 liters of water per day and are effective strainers of particles as small as 1 μm or less.

Live and dead oysters are probable reservoirs of infective particles. Barrow (1965) reported direct transmission of the pathogen *Minchinia pickfordae* among snails. Both the plasmodial and spore stages of MSX are probably infective. The scarcity of spores of *M. nelsoni* leads one to believe that another host transmits MSX. In contrast, *M. costalis* produces many spores. For both oyster pathogens, the infection period occurs during outbreaks of major oyster population decline. This period is most striking for *M. costalis* infection by which and the resulting oyster mortality is usually limited to a period of 6 to 8 weeks or less.

Apparently, the method of transmission among oysters is by water currents, despite the attendant dilution. Plasmodia expelled by oysters during regression of MSX infections appear to be moribund, leaving primarily disintegrating gapers or dying oysters as the major source of infective particles. The regularity of the incidence of haplosporidan diseases seems to require a simple, direct method of transmission.

Patterns of Infection. Early summer is a major period of oyster infection with both *M. nelsoni* and *M. costalis*, although prolonged mortalities from MSX through October are known. *Marteilia refringens* (Aber disease) also infects oysters in early summer but mortality occurs in the late summer of the following year (Balouet, 1979). Cold summer waters (<20°C) of western Europe probably inhibit rapid development of this pathogen; the closely related pathogen, *M. sydneyi* of Australia, kills soon after infection. Aber disease is said to regress in winter and reappear in spring without evidence of new infections.

The site of MSX infections is usually in the gill epiethelium. Early plasmodia are often located in the food groove at the base of the lamellae. These localized infections do not elicit much response by oysters. The epithelium along one side of a gill may be infected by hundreds of plasmodia over a considerable length without any subepithelial infection. The opposing gill face often does not show infection. Either numerous infective particles invade the epithelium, or the infection somehow spreads along the epithelial border without entering the blood sinuses beneath the basement membrane. No known mechanism by which plasmodia can be distributed along the epithelial layers of tissues has been reported.

The titer for an effective dose of infective particles is unknown but presumed to be low because of wide dispersal of MSX disease in tidal waters. If infective particles released from dying oysters infect oysters tens of miles away in one season, infectivity must be very high, that is, produced by a few infective particles. The highly localized infections in gill epithelia of oysters commonly observed, and the long incubation periods that indicate low-level infections seem to confirm this contention. Within endemic areas MSX exhibits wide distribution regardless of the distance from experimental lots and populations of natural or seeded beds. Approximately equal infection levels at all sites may be explained by high infectivity requiring only a few infective particles to establish the diseases. MSX exhibits both high infectivity and high pathogenicity which makes it a very serious disease of oysters.

Incubation Periods for M. nelsoni and M. costalis. Oysters exposed to MSX in early summer (late May to mid-July) are rapidly infected. These earliest infections become clinical about 4 weeks after transplantation of oysters to infested areas. Death of the oysters begin 6-8 weeks after their first exposure (usually about 1 August). Mortality continues through the late summer, then declines in November and December. A late winter kill (March) is followed by renewed mortality in June and July of the second year (Andrews and Frierman, 1974).

Oysters transplanted into an MSX-prevalent area between 1 August and 1 November acquire subclinical infections that remain latent until May-June of the following year (Andrews and Frierman, 1974). This long incubation period has been attributed to minor localized infections that are overlooked in the examination of tissue sections (Myhre, 1972). The oysters exhibit little hemocytic reaction in the fall to these localized infections. Why infections acquired during August in warm temperatures should remain quiescent for 8 to 10 months and then suddenly develop rapidly in the cool waters of May is unknown. Presumably, the intensity of the initial infection is a factor. In some years when mortality rates are high late summer infection becomes clinical and systemic as early December. Once the infection becomes systemic, there is a strong hemocytic response in the tissues. No evidence of new infections being acquired between November and early May has been found. All lots of oysters exposed during that period showed earliest infections in late June or early July at the same time as May-June importations. In some years, no late summer infections are acquired after 1 August.

M. costalis always exhibits an incubation period of 8 to 10 months. Localized infections are rarely found in the gills and it is believed that digestive tubule epithelia are the primary sites of infection. The early, uninucleate stage is difficult to find in tubule epithelia and it does not invade the connective tissues or develop into plasmodia until spring (Andrews and Castagna, 1978).

Clinical Signs. Oysters with systemic infections of MSX stop feeding and soon become low in glycogen reserves (Andrews, 1961). Nearly all macroscopic signs described by Farley (1968) for Delaware Bay disease (MSX) are generalized indications of sick or diseased oysters. It is necessary to make diagnoses by examination of stained tissue sections for plasmodia. (The regression of the flat right value of the shell occurs in oysters with prolonged sickness from many causes.) Discoloration is usually confined to a darkening of the tissue as the digestive gland becomes visible through the transparent tissues. In rare cases scabby brown deposits of conchiolin on the inner surface of shells result from abcesses on the mantle caused specifically by MSX (Farley, 1968). These conspicuous shell deformations rarely appeared in the mid-1960s when MSX was widespread and exhibiting high prevalences in previously unexposed populations. The disease spread northward in the Chesapeake Bay among highly susceptible oyster populations during a period of high salinities caused by drought. The scabby spots have not been seen in recent years.

There are no specific macroscopic signs of *M. costalis* infections. Infected oysters are conspicuous by the absence of new bills during the spring growth period of the shell. The tissues are not adequately discolored by sporulation to identify infected oysters or gapers. Van Banning (1977) reported a discoloration of tissues by sporocysts of *Minchinia amoricana.*

Prevalence. Occurrence of infections in samples of live oysters at any point in time can define the prevalence of disease. Incidence of new infections for a fixed period of time is more difficult to measure.

M. nelsoni is more virulent than *M. costalis* in the Chesapeake Bay area. Most seed oysters exposed to *M. nelsoni* are transplanted from areas not previously exposed and are susceptible to the pathogen. In contrast, reproduction of young oysters on the seaward-side of the eastern shore of Virginia occurs in an endemic area and, therefore, all oysters are exposed to *M. costalis* during May-June of each year.

Prevalences of MSX in endemic areas usually reach levels of 40 to 50% in late summer, decline in late fall and winter, followed by new highs of 50 to 60% in May that result from late summer infections (Andrews, 1966). Haskin (1972) reported higher prevalences in the Delaware Bay. A few highly susceptible lots of oysters have exhibited prevalences of 90% during the summer in Virginia (Andrews, 1979).

M. Costalis exhibits prevalences of 30 to 40% infections in most years. The prevalences for 1979 shown in Figure 2 are exceptionally low, especially in Chincoteague Bay, and only 54% of 39 gapers had the disease. Unlike the irregular appearance of MSX infections, all diagnosed cases become clinical rather suddenly during April-May just prior to death. In Virginia, sampling of oysters to establish prevalences and morbidity rates should be done about 15 May in most years. For the disease caused by *M. costalis* (seaside disease) it is relatively easy to relate morbidity to mortality, since they occur in quick succession. Sporulation occurs regularly in nearly all oysters and causes death within a short period (ca. 1 month).

In the second year of infection, both pathogens affect nearly the same proportion of the surviving population. The mortality data indicate that minor kills by *M. costalis* continue for as long as 5 years in the same population.

Morbidity and Mortality. Prevalences of MSX infections are not as high as would be expected based on annual mortalities caused by the disease. Infections appear over 5 months of the year and deaths occur concurrently for 3 or 4 of these months. Therefore, infected oysters are being continuously added and lost through the summer and fall. Late summer infections result in another mortality period in June and July of the following year. There are two rather distinct infection periods that result in overlapping mortality periods within a year. Even assuming no regression of infected cases, it is difficult to estimate morbidity rates and the ratio of deaths to cases of infection. In Delaware Bay, infections seem more prevalent than in the Chesapeake Bay because of the higher prevalences and quicker deaths of infected individuals (Haskin, 1972). However, after 2 or 3 years of exposure to MSX, 80 to 90% of most susceptible lots of oysters are killed by the disease in both the Delaware and Chesapeake Bays. In the Chesapeake Bay, annual mortalities usually exceed 50% during the first year; in the second year losses are nearly as high. These rates have been measured using transplanted James River oysters over a period of 20 years (Andrews and Frierman, 1974).

M. costalis has a well defined annual life-cycle with a June-July period of infectivity and a regular May-June mortality the following year. Morbidity rates are nearly equivalent to prevalence rates and are easily compared to mortality rates a month later. A few light infections regress before sporulation occurs which reduces slightly the ratio of deaths to cases of infection.

For both diseases, the morbidity rate or ratio of deaths to cases is high -- probably 80 to 90% within suitable salinity ranges for the pathogens. However, in low salinities (<10°/oo), most infections regress in early fall (September). Regression of MSX in the seed oysters from the James River occurs about 1 May each year when the salinities are low. This regression in spring occurs in areas with late summer salinities of 15-18°/oo, permitting reinfections in late summer and fall. These late-summer reinfections remain subclinical until just before oysters expel them in May. If oysters with subclinical infections are transplanted into waters of higher salinity, the pathogen develops and eventually causes death. In the James River seed area, there is no feasible way of diagnosing subclinical infections, therefore, transplanting these oysters to salty waters is not advisable. In most years there are no infections in the seed area except on the downstream edge.

Age Groups Affected. Although MSX occurs in oysters of all ages, including spat only 2 months old (Andrews, 1979), spat are not usually killed by MSX in the Chesapeake Bay. Presumably, the dose is inadequate or spat remove infections before they become systemic (Myhre, 1972). When the oysters have reached a length of at least 50 mm by the second summer, they are subjected to intensive selection by MSX.

The occasional MSX infections of spat transmitted by susceptible parents indicate that all spat are exposed to infective particles when feeding. Apparently, most young oysters are able to phagocytize or discard early stage pathogens as rapidly as they are collected, thereby usually avoiding infections.

M. costalis is less likely to infect young oysters and thus mortality is highest after 2 years of age, i.e., when oysters are market size and past the usual marketing age. Spat are not exposed to seaside disease until they are nearly 1-year-old because the short annual period of infectivity occurs before spatfall. Holding oysters an extra year to obtain large raw bar oysters is risky because of the potential for seriously reduced yields from infection.

Both *Marteilia refringens* and *M. sydneyi* are reported to kill young oysters in the first year of culture (Alderman, 1979; Wolf, 1979).

Susceptibility and Resistance. The regular occurrence and rather uniform distribution of MSX infections over a period of 20 years has provided a reliable test of the relative resistance of oysters to these pathogens. None of the imported populations from New England to South Carolina are resistant upon first exposure. After intensive selection over many years, native oysters of Delaware Bay showed relatively high resistance (Haskin and Ford, 1979). Nearly all populations in Delaware Bay were subjected to selection by the disease. Delaware Bay contains few low salinity beds. Artificial breeding in the laboratory yielded even higher resistance in progeny after 3 years of selection within each brood stock. By selective breeding and seeding, resistance was attained in the James River, Long Island, and several other stocks that originally were highly susceptible to MSX.

In the Chesapeake Bay large stocks of susceptible oysters continue to reproduce without incidence of disease. These stocks in low salinity sanctuaries continue to provide larvae for spatfalls in the lower rivers where MSX is prevalent. Therefore, resistant brood stocks have been slow to develop in high salinity areas where MSX is endemic. Some resistant populations have appeared in the York River where a low salinity sanctuary is essentially absent. A bed of 3-year-old oysters was harvested in 1978 from a natural set on a State shell planting. Very few deaths had occurred and a sample lot, held in a tray for a year in an MSX area, had negligible deaths.

Laboratory-bred stocks from survivors of the MSX epizootic mortality in Mobjack Bay in 1959-60 have exhibited strong resistance since 1964. Control lots bred from susceptible James River seed oysters were held each year at Gloucester Point. They experienced mortalities of about 50% the first year with slightly less mortality the second year. Over 100 lots bred from surviving parents since 1964 have exhibited a high level of resistance with only about 10% mortality each year, including unknown causes of death. The oysters are healthy and seldom have prevalences of MSX higher than 10 to 15%.

The most susceptible oysters were found in the upper Potomac River in salinities below 15°/oo where MSX was never observed and presumably exposure has never occurred (Andrews, 1968; Valiulis, 1972). Even these highly susceptible lots, which exhibit prevalences and mortalities well above average, provided a few survivors whose progeny exhibited greatly increased resistance.

The general or specific mechanisms of resistance are unknown. All native populations of oysters include some resistant individuals that could be useful in selective breeding programs.

Comparison of Three Chesapeake Bay Oyster Diseases. The newly renamed apicomplexan pathogen *Perkinsus marinus* was formerly believed to be a fungus (*Dermocystidium*) and then a labyrinthulid (Mackin and Ray, 1966). *P. marinus* is distributed from Delaware Bay to the Gulf of Mexico. Pathogenicity is highly dependent on warm temperatures; at temperatures below 20°C oysters rapidly expel the pathogen. Summer infections reach peaks in September and October resulting in serious mortalities in the fall if temperatures remain high. If cool temperatures prevail in early fall, development of infection is inhibited and death is slow. By mid-December winter dormancy begins and most oysters have expelled the disease causing organism. A few subclinical infections are carried through the winter and begin a new season for the disease from June through September. Changes in levels of fall temperatures and winter survival of diseased oysters cause large annual differences in enzootic mortalities and in the rate of pathogen dispersal. Infections are acquired from infective particles released from dying and infected oysters in close proximity to healthy oysters. Changes in density of oyster populations dramatically affect the incidence of *P. marinus* and its distribution. Therefore, *P. marinus* tends to be localized in distribution and the mortality levels caused by it are irregular.

In contrast to *P. marinus*, the two haplosporidan pathogens are widely distributed in waters with appropriate salinities, and dispersal is rapid and extensive with a regular incidence of the disease. All three disease causing organisms have persisted without evident changes of virulence through 20 to 30 years of observation. *P. marinus* declined in abundance in the lower Chesapeake Bay after 1960 because of the elimination of oyster seeding and the reduction of populations on natural beds following the MSX catastrophy (Andrews and Wood, 1967).

All three disease causing organisms are prevalent in the warm season and oysters acquire infections by filtering infective particles from the water during food collection. None of the pathogens are active in the cold season, although the haplosporidans persist through the winter without evident declines in prevalence caused by low temperatures. All are systemic diseases that disrupt growth and reproduction. Alternate hosts have not been demonstrated for them. There is no proof that these infectious agents have breeding places outside oysters. Presumably, diseased cadavers or gapers are the primary source of infective particles. It is not known how long infective particles may survive outside the oyster. From 10 to 1000 infective particles of *P. marinus* are required to achieve infections in oysters (Mackin 1961; Valiulis, 1972), but crowding results in nearly 100% infection of oysters in a bed or tray. High levels of infection occur when *P. marinus*-infected individuals over-winter in a population to produce early infections in June. Subsequent release of infective particles initiate a second round of disease during the warm season.

Density of oysters appears to have no effect on the two haplosporidans since prevalence and morbidity tend to be similar each year (30 to 60%). Prevalences are consistent throughout the lower Chesapeake Bay or on the seaward-side of the eastern shore each year. This characteristic level of infection continues into the second year in the same populations of oysters.

Knowledge of age susceptibility of oysters to disease organisms is not well understood. Young oysters tend to escape the diseases in nature. Rapid development of MSX resistant stocks and selection by MSX tend to obscure differences in susceptibility of various age classes of oysters. The long five month infective period insures that current-year spat are exposed to MSX in their first months after setting. With rare exceptions (Andrews, 1979), no mortality from MSX occurs during the first 10 months of life, but spat may carry latent infections from late summer into the spring (Myhre, 1972). Infections of spat were found more common in Delaware Bay than in the Chesapeake Bay (Myhre, 1972).

Although spat are not usually infected by *P. marinus* in natural waters, grouping of oysters of several ages in close proximity provides infective dosages or demonstrated by Andrews and Hewatt (1957) in aquarium infections which persist in natural waters. One assumes that the low infectivity of *P. marinus* for young oysters is related to dosage and their rates of collection and explusion of infective particles. A relatively high dosage is required for *P. marinus* to infect oysters. Since plasmodia of haplosporidans are not as readily phagocytized as the plasmodia of *P. marinus*, infections by pathogens of this group may be achieved in young oysters more regularly.

Selection of stocks resistant to *P. marinus* has not been effective in Virginia although Valiulis (1972) claimed that MSX-resistant strains of oysters showed some resistance to the apicomplexan parasite. I found that surviving oysters dredged from Mobjack Bay after 5 years of MSX selection continued to die from *P. marinus* disease. An intensive effort to exclude the disease from laboratory-bred MSX-resistant oysters by isolation has been moderately successful. Once established in MSX-resistant oysters in isolated trays, *P. marinus* continued to cause infections and deaths for the duration of the tests. The method of infection that requires close proximity to dying oysters and large dosages almost prohibits selection of strains that are resistant to *P. marinus*. It is impossible to space oysters on natural bottoms in commercial operations to avoid intensive exposure. However, the course of *P. marinus* disease may be slowed or altered by races with generalized resistance mechanisms. Any delay in the timing of infections and deaths tends to reduce epizootic mortalities and limit secondary rounds of infection. Also, declining temperatures in the fall enable most oysters to recover from the disease before winter.

All races of oysters from Long Island to the Chesapeake Bay have shown increased resistance to MSX after two or three generations of selective breeding. Experimental populations showed 80 to 90% mortality before brood stocks were selected. In Virginia stocks, a rather high level of resistance has been achieved after intensive selection in one generation. Tests of MSX-resistant oysters for resistance to *M. costalis* have been inconclusive because of interference by other diseases.

IV. TRANSMISSION

The method of transmission of haplosporidian pathogens remains a mystery. No one has succeeded in producing laboratory or experimental infections with spores or tissue extracts of diseased oysters with any haplosporidan. In contrast, experimental infection with *P. marinus* is easily accomplished. Unfortunately, *M. nelsoni* rarely sporulates and only casual infection attempts have been made with spores of this species. No one has reported a spore case clinging to the gills of oysters during examination of tens of thousands of diseased specimens. Presumably, free sporoplasms and plasmodia represent the infective stages. Well over a dozen species of haplosporidan pathogens representing the genera *Minchinia, Haplosporidium, Urosporidium,* and *Marteilia* exhibit sporulation routinely (Farley, 1967; Sprague, 1970, 1971, 1979; Kern, 1976; Perkins, 1979). At least five or six occur in bivalve shellfish. The scarcity of *M. nelsoni* spores cannot be used as evidence against direct transmission from oyster to oyster in most haplosporidans. Prior to the earliest attempts at experimental infections with *M. nelsoni* (Couch et al., 1966), only two haplosporidians had been tried in direct transmission experiments and one species was reported to be successfully transmitted (Barrow, 1965). *Nephridiophaga blattellae* in German cockroaches (Woolever, 1966) was readily infected *per os* with spores. However, Sprague (1979) has rejected the family as a member of the Haplosporida.

The method of transmission of haplosporidan pathogens in flatworms and nematodes parasitic in bivalve molluscs and crabs is equally puzzling (Perkins, 1971). It is unknown whether or not flatworms carry transovarian infections or acquire infections while parasitic in the host. One can only assume that infections are acquired during the short periods in which the free-living stage exists. *Bucephalus cuculus* castrates oysters and occupies space in the connective tissues adjacent to the digestive tubules where infective stages could be ingested. The method by which nematodes isolated from the tissues of surf clams become infected with *Urosporidium spisuli* is not well understood (Perkins, 1975). Perkins states that it is unlikely this haplosporidan is transmitted from nematode to nematode. But what advantage is there to the life cycle of the parasite, whether the infective stage originates from the nematode or another host, if it must transgress oyster defenses to get to the nematode? Oysters serving as an alternate host, tolerate the porosporan *Nematopsis* and encyst temporarily in the tissues (Feng, 1958), but to pass through the host tissues without parasitic capability seems unlikely.

The rapidity of initial dispersal of MSX in the Chesapeake and Delaware Bays illustrates the effectiveness of the method of transmission. In one year the disease spread to the limits of its salinity tolerance in both bays. In the Chesapeake Bay, man did not spread the disease by transplantation of oysters since the primary seed area was free of the disease in 1959-60 and has remained free during most years. If an indigenous alternate host was involved, it spread the disease at a fast rate, which would include most small resident invertebrates and scavengers associated with oyster beds, but no necessarily planktonic species. No new exotic species of invertebrates were discovered in Chesapeake Bay before or after the appearance of the Delaware Bay disease.

When MSX first appeared in the Chesapeake Bay in 1959, there were several million bushels of oysters planted in the lower part of the Bay and in a few public grounds. These populations declined drastically in a 2-year period and have never been replaced naturally or artificially. Yet, the Delaware Bay disease has continued to kill transplanted susceptible oysters in trays and on trial plots at approximately the same rate that has existed for 20 years. One might have expected a decline in the abundance of infective particles comparable to the decline in oysters, if dying oysters were the source. In fact, the highest prevalence and greatest mortality occurred in the dry years of 1964-66 and after the decline in the oyster populations.

Higher salinities in the mid-1960's allowed exposure of susceptible oyster populations northward in the Chesapeake Bay to the disease. The years of highest mortalities are usually those in which deaths occur early. In some years mortalities commenced a month in advance of the average date of 1 August. Thus, low salinities in the early infective period of mid-May through June may be unfavorable to the disease. During the infection period in wet years, salinities may be below the 15°/oo believed to be necessary for infection to occur. It has been shown that in areas of marginal salinities (<15°/oo) clinical infections in the James River were delayed in appearance until at least November (Andrews, 1968).

The May-June growing period provides optimal conditions for oyster growth and storage of glycogen necessary for gametogenesis. Weekly increments of live oyster weights are greatest in these months. High water temperatures in the summer alter the species composition and quantity of plankton crops and conditions become less favorable for oyster pumping. The May-June period is also the prime time for oyster infections to occur. August and September are prime months for mortality from MSX disease.

Without information on the infective stage, there is no solid basis for a transmission theory of haplosporidan diseases. However, all evidence points to an annual life cycle with a June-July period of sporulation. Sporulation is usually associated with deaths but for MSX and the *Marteilia* diseases, it may not cause immediate deaths. MSX is an exceptional disease-causing species that is highly virulent in Virginia oysters, and therefore, often surprisingly light plasmodial infections cause death. Disease control must depend upon development of resistant strains of oysters. Some manipulation of oysters using low-salinity areas and proper timing of transplantations to avoid latent infections is possible. However, susceptible oysters are infected and killed so quickly (2-4 months) that high-salinity areas cannot be used effectively for culture without resistant oysters. A solution to the disease problem in Chesapeake Bay may lie in substituting resistant races in the lower part of the Bay for the susceptible ones still breeding in low-salinity areas without seriously altering the genetic diversity and adaptations of native oysters. Unfortunately, large hatcheries to produce resistant seed oysters from selected brood stocks are not presently available.

V. CONCLUSIONS

Haplosporidan pathogens of oysters have been found in the temperate zones of four continents, and four species in the temperate zones of three of these continents have caused serious mortalities. Two species, *Minchinia nelsoni* and *M. costalis*, have persisted along the middle Atlantic Coast of North America for two decades without significant changes in the epizootiology of the diseases. The Delaware Bay disease (*M. nelsoni*) causes mortalities of 30 to 60% annually in the Chesapeake Bay among susceptible James River seed oysters. Populations of oysters in this area have declined greatly, but the disease continues to kill oysters at the same high rates. Therefore, oysters are not planted in areas of moderate- to high-salinity Chesapeake Bay waters where Delaware Bay disease is now endemic. Seaside disease (*M. costalis*) is restricted to high-salinity waters along the Atlantic coastline and appears to have been endemic for many years.

Both diseases are highly infective in their endemic areas with rapid dispersal of infection and high ratios of deaths to disease cases. The sources and stages of infective agents are not known. The failure to achieve experimental infections precludes a thorough description of the life cycles. Both *M. nelsoni* and *M. costalis* kill primarily during the warm season but persist throughout the cold season, often as subclinical or localized infections. Alternate hosts have not been demonstrated for any haplosporidan. Most species sporulate regularly and kill their hosts; new infections occur during this period of mortality. In addition to bivalve hosts, haplosporidans are parasites of chitons, nemerteans, annelids, and crustaceans. They are also hyperparasites of flatworms and nematodes parasitic in bivalve molluscs and crustaceans (Sprague, 1970).

Approximately a decade ago, two closely related pathogens, *Marteilia refringens* and *M. sydneyi*, appeared in oysters of Western Europe (*Ostrea edulis*) and Australia (*Crassostrea commercialis*), respectively. These pathogens caused serious mortalities. Unlike the two *Minchinia* species, which cause systemic infections, these haplosporidans parasitize primarily the digestive system. *Marteilia* spp. appear to cause starvation of oysters at lower intensities of infection than found in *Crassostrea virginica* in North America.

The epizootiology of the two haplosporidan diseases in the Chesapeake Bay is compared with the disease caused by *Perkinsus marinus*, now assigned to the new subphylum Apicomplexa. This well known disease spreads directly to oysters by exposure to disintegrating infected oyster cadavers or gapers. Unlike the haplosporidans, which are best controlled by development of resistant strains of oysters, *P. marinus* can be controlled by avoidance of infected lots and by cleaning oysters from old beds to eliminate sources of infection. In the Chesapeake Bay, sanctuaries from all three diseases are provided by areas with salinities below 15°/oo. Most oysters are now grown in the low-salinity areas. Most oyster-growing regions around the world are high-salinity areas and lack these sanctuaries.

VI. REFERENCES

Alderman, D. J. (1979). Epizootiology of *Marteilia refringens* in Europe. *Marine Fish. Review*, 41, 67-69.

Andrews, J. D. (1961). Measurement of shell growth in oysters by weighing under water. *Proc. Natl. Shellfish Assoc.*, 52, 1-11.

Andrews, J. D. (1966). Oyster Mortality studies in Virginia. V. Epizootiology of MSX (*Minchinia nelsoni*), a protistan pathogen of oysters. *Ecology*, 47, 19-31.

Andrews, J. D. (1968). Oyster mortality studies in Virginia. VII. Review of epizootiology and origin of *Minchinia nelsoni*. *Proc. Natl. Shellfish Assoc.*, 58, 23-36.

Andrews, J. D. (1979). Oyster diseases in Chesapeake Bay. *Marine Fish. Review*, 41, 45-53.

Andrews, J. D. and Hewatt, W. G. (1957). Oyster mortality studies in Virginia. II. The fungus disease caused by *Dermocystidium marinum* (now *Perkinsus marinus*) in oysters of Chesapeake Bay. *Ecol. Monogr.*, 27, 1-26.

Andrews, J. D. and Wood, J. L. (1967). Oyster mortality studies in Virginia. VI. History and distribution of *Minchinia nelsoni*, a pathogen of oysters in Virginia. *Chesapeake Sci.*, 8, 1-13.

Andrews, J. D. and Frierman, M. (1974). Epizootiology of *Minchinia nelsoni* in susceptible wild oysters in Virginia, 1959-1971. *J. Invertebr. Pathol.*, 24, 127-140.

Andrews, J. D. and Castagna, M. (1978). Epizootiology of *Minchinia costalis* in susceptible oysters in Seaside Bays of Virginia's Eastern Shore, 1959-1976. *J. Invertebr. Pathol.*, 32, 124-138.

Balouet, G. (1979). *Marteilia refringens*-- considerations of the life cycle and development of Abers disease in *Ostrea edulis*. *Marine Fish. Review*, 41, 64-66.

Barrow, J. H., Jr. (1965). Observations on *Minchinia pickfordae* (Barrow, 1961) found in snails of the great lakes region. *Trans. Am. Microsc. Soc.*, 84, 587-593.

Couch, J. A., Farley, C. A., and Rosenfield, A. (1966). Sporulation of *Minchinia nelsoni* (Haplosporida, Haplosporididae) in the American oyster *Crassostrea virginica*. *Science*, 153, 1529-1531.

Farley, C. A. (1967). A proposed life cycle of *Minchinia nelsoni* (Haplosporida: Haplosporidiidae) in the American oyster *Crassostrea virginica*. *J. Protozool.*, 14, 616-625.

Farley, C. A. (1968). *Minchinia nelsoni* (Haplosporida) disease syndrome in the American oyster *Crassostrea virginica*. *J. Protozool.*, 15, 585-599.

Feng, S. Y. (1958). Observations on distribution and elimination of spores of *Nematopsis ostrearum* in oysters. *Proc. Natl. Shellfish. Assoc.*, 48, 162-173.

Frost, W. H. (1941). Epidemiology. *In* "Papers of Wade Hampton Frost, M. D., A Contribution to Epidemiological Method." The Commonwealth Fund, Oxford Univ. Press, London, pp. 493-542.

Haskin, H. H. (1972). Disease resistant oyster program - Delaware Bay 1965-1972. Unpubl. report to Natl. Marine Fish. Service.

Haskin, H. H., Stauber, L. A., and Mackin, J. G. (1966). *Minchinia nelsoni* n. sp. (Haplosporida: Haplosporidiidae) causative agent of the Delaware Bay oyster epizootic. *Science*, 153, 1444-1416.

Haskin, H. H. and Ford, S. E. (1979). Development of resistance to *Minchinia nelsoni* (MSX) mortality in laboratory-reared and native oyster stocks in Delaware Bay. *Marine Fish. Review*, 41, 54-63.

Hillman, R. E. (1979). Occurrence of *Minchinia* sp. in species of the molluscan borer, Teredo. *Marine Fish. Review*, 41, 21-24.

Kern, F. (1976). Sporulation of *Minchinia* sp. (Haplosporidiidae) in the Pacific oyster *Crassostrea gigas* (Thunberg) from the Republic of Korea. *J. Protozoology*, 23, 498-500.

Mackin, J. G. (1961). Mortalities of oyster. *Proc. Natl. Shellfish Assoc.*, 51, 21-40.

Mackin, J. G. and Loesch, H. (1955). A haplosporidian hyperparasite of oysters. *Proc. Natl. Shellfish Assoc.*, 45, 182-183.

Mackin, J. G. and Ray, S. M. (1966). The taxonomic relationships of *Dermocystidium marinum* Mackin, Owen, and Collier. *J. Invertebr. Pathol.*, 8, 544-545.

Marteil. L. (1968). La "maladie des branchies". *ICES, Shellfish and Benthos Comm.* Doc. CM 1968/K:5.

Marteil, L. (1976). La Conchyliculture Française. *Rev. Trav. Inst. Peches Marit.*, 40, 149-346.

Myhre, J. L. (1972). *Minchinia nelsoni* (MSX) infections in resistant and susceptible oyster stocks. *In* "Haskin Unpubl. Report to Natl. Marine Fish. Service."

Perkins, F. O. (1968). Fine structure of the oyster pathogen *Minchinia nelsoni* (Haplosporida: Haplosporidiidae). *J. Invertebr. Pathol.*, 10, 287-305.

Perkins, F. O. (1971). Sporulation in the trematode hyperparasite *Urosporidium crescens* Deturk, 1940 (Haplosporida: Haplosporidiidae): an electron microscope study. *J. Parasitol.*, 57, 897-920.

Perkins, F. O. (1975). The hyperparasite *Urosporidium spisuli* sp. n. (Haplosporea), and its effect on the surf clam industry. *J. Parasitol.*, 61, 944-949.

Perkins, F. O. (1979). Cell structure of shellfish pathogens and hyperparasites in the genera *Minchinia, Urosporidium, Haplosporidium* and *Marteilia*-- taxonomic implications. *Marine Fish. Review*, 41, 25-37.

Rosenfield, A., Buchanan, L., and Chapman, G. B. (1969). Comparison of the fine structure of spores of three species of *Minchinia* (Haplosporida: Haplosporidiidae). *J. Parasitol.*, 55, 921-941.

Sindermann, C. J. (1976). Oyster mortalities and their control. *FAO Technical Conf. on Aquaculture*, Kyoto, Japan.

Sprague, V. (1970). Some protozoan parasites and hyperparasites in marine bivalve molluscs. *In* "A Symposium on diseases of fishes and shellfishes" (K. Snieszko, ed.), *Spec. Publ. 5, Am. Fish. Soc.*, Washington, D.C.

Sprague, V. (1971). Diseases of oysters. *Ann. Rev. Microbiol.*, 25, 211-230.

Sprague, V. (1979). Classification of the Haplosporidia. *Marine Fish. Review*, 41, 40-44.

Taylor, R. I. (1966). *Haplosporidium tumefacientis* sp. n., the etiologic agent of disease of the California sea mussel, *Mytilus californianus* Conrad. *J. Invertebr. Pathol.*, 8, 109-121.

Valiulis, G. (1972). Comparison of resistance to *Labyrinthomyxa marina* (now *Perkinsus marinus*) with resistance to *Minchinia nelsoni* in *Crassostrea virginica In* "Haskin unpubl. report to Natl. Marine Fish. Service".

Van Banning, P. (1977). *Minchinia amoricana* sp. nov. (Haplosporida), a parasite of the European flat oyster, *Ostrea edulis*. *J. Invertebr. Pathol.*, 30, 199-206.

Van Banning, P. (1979). Haplosporidian diseases of imported oysters, *Ostrea edulis*, in Dutch estuaries. *Marine Fish. Review*, 136, 8-18.

Wolf, P. (1979). Life cylce and ecology of *Marteilia sydneyi* in the Australian oyster *Crassostrea commercialis*. *Marine Fish. Review*, 41, 70-72.

Wood, J. L. and Andrews, J. D. (1962). *Haplosporidium costale* (Sporozoa) associated with a disease of Virginia oysters. *Science*, 136, 710-711.

Woolever, P. (1966). Life history and electron microscopy of a hapsporidian, *Nephridiophaga blattellae* (Crawley) n. comb., in the malphighian tubules of the German cockroach, *Blattella germanica* (L). *J. Protozool.*, 13, 622-642.

INDEX

MIX
Papier aus verantwortungsvollen Quellen
Paper from responsible sources
FSC® C105338

If you have any concerns about our products,
you can contact us on
ProductSafety@springernature.com

In case Publisher is established outside the EU,
the EU authorized representative is:
Springer Nature Customer Service Center GmbH
Europaplatz 3, 69115 Heidelberg, Germany

Printed by Libri Plureos GmbH
in Hamburg, Germany